Essener Beiträge zur Mathematikdidaktik

Series Editors

Bärbel Barzel, Fakultät für Mathematik, Universität Duisburg-Essen, Essen, Germany

Andreas Büchter, Fakultät für Mathematik, Universität Duisburg-Essen, Essen, Nordrhein-Westfalen, Germany

Florian Schacht, Fakultät für Mathematik, Universität Duisburg-Essen, Essen, Germany

Petra Scherer, Fakultät für Mathematik, Universität Duisburg-Essen, Essen, Nordrhein-Westfalen, Germany

In der Reihe werden ausgewählte exzellente Forschungsarbeiten publiziert, die das breite Spektrum der mathematikdidaktischen Forschung am Hochschulstandort Essen repräsentieren. Dieses umfasst qualitative und quantitative empirische Studien zum Lehren und Lernen von Mathematik vom Elementarbereich über die verschiedenen Schulstufen bis zur Hochschule sowie zur Lehrerbildung. Die publizierten Arbeiten sind Beiträge zur mathematikdidaktischen Grundlagen- und Entwicklungsforschung und zum Teil interdisziplinär angelegt. In der Reihe erscheinen neben Qualifikationsarbeiten auch Publikationen aus weiteren Essener Forschungsprojekten.

More information about this series at http://www.springer.com/series/13887

Lisa Göbel

Technology-Assisted Guided Discovery to Support Learning

Investigating the Role of Parameters in Quadratic Functions

With a Preface by Prof. Dr. Bärbel Barzel

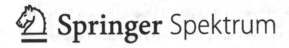 Springer Spektrum

Lisa Göbel
Didaktik der Mathematik
Universität Duisburg-Essen
Essen, Germany

Dissertation der Universität Duisburg-Essen, 2020
Von der Fakultät für Mathematik der Universität Duisburg Essen unter dem Titel „Power of Speed or Discovery by Slowness – Technology-assisted Guided Discovery to Investigate the Role of Parameters in Quadratic Functions" genehmigte Dissertation zur Erlangung des Doktorgrades "Dr. rer. nat."
Datum der mündlichen Prüfung: 31. August 2020
Gutachterinnen: Prof. Dr. Bärbel Barzel, Dr. Lynda Ball

ISSN 2509-3169   ISSN 2509-3177 (electronic)
Essener Beiträge zur Mathematikdidaktik
ISBN 978-3-658-32636-4   ISBN 978-3-658-32637-1   (eBook)
https://doi.org/10.1007/978-3-658-32637-1

Responsible Editor: Marija Kojic
This Springer Spektrum imprint is published by the registered company Springer Fachmedien Wiesbaden GmbH part of Springer Nature.
The registered company address is: Abraham-Lincoln-Str. 46, 65189 Wiesbaden, Germany

Preface

Digitization is currently the central challenge in the educational landscape. In mathematics education digitization is not only about digital media for enhancing communication and presentation but first of all the meaningful integration of math specific digital tools like spreadsheets or function plotters.

These tools play an important role in learning mathematics and are firmly anchored in all curricula on secondary level. Particularly quantitative research has shown that tools like function plotters can have a positive influence on learning and understanding mathematics. In detail, however, it is still unclear, how exactly this influence works and what role various technical features such as sliders or drag mode can play.

The work of Lisa Göbel can be located here. Her work is based on research findings that the potential of the use of media can unfold, especially in learning environments that are committed to exploratory learning. Therefore, she has designed a learning environment for exploring the meaning of parameters in quadratic functions, which follows the structure of a guided discovery approach. In this context, she has carried out a comparative study with three experimental groups and one control group in 14 classes of grade 9. All four groups follow the same work order, the control group without using technology and the three experimental groups with technology in different technical variations—one group used a standard function plotter, the two others prepared files to change graphs either via drag mode or via slider.

Lisa Göbel followed a mixed method design with quantitative and qualitative approaches aiming in getting a better insight view which role technology with the different features play. The investigations conclude in clear results. The designed guided discovery approach is successful, with a clear advantage for the group that used dynamic tools (drag mode or sliders), where drag mode is best suited for the vertical changes and sliders better to understand the horizontal movements and the influence of parameter a. Beside this very differentiated picture of the potentials, further details showed the benefit of the whole orchestration of the learning environment as a guided discovery approach. The work gives insight into the interplay between the different levels of communication with and of technology, the use of the different technical tools and the content learning about the meaning of parameters in quadratic functions.

The design of the learning environment as well as the research of the students' work is done by Lisa Göbel on a high level and provides valuable insights for the further scientific discourse in the area of the use of digital mathematical tools which enrich the mathematics didactic communication about the role of technology when learning mathematics.

Essen   Prof. Dr. Bärbel Barzel
Oktober 2020

Acknowledgements

This dissertation marks the completion of over five years of work and many people helped bring this process to a successful conclusion, both through professional and personal support.

First and foremost, I would like to thank my supervisor Prof. Dr. Bärbel Barzel. From the beginning she supported me in my endeavors and was also there for me in difficult times with encouraging words and constructive feedback. Her enthusiasm for imparting knowledge has left a lasting impression on me and is something I aspire to achieve.

A sincere thanks to my second supervisor Dr. Lynda Ball, who gave important advice and welcomed me to the Melbourne Graduate School of Education for an extended stay. She supported my dissertation with her international expertise and offered an insight into the mathematics education community of Australia.

Thank you to Prof. Dr. Petra Wittbold for chairing my examination committee and also for supporting me during my teaching degree.

I would like to thank the school principles, teachers and students involved without whom the thesis would not have been possible. The coding of the data would not have been possible without the support of Hülya Duman, Sasha Hönecke, Joscha Klein-Altstedde, Serap Kurtul, Susan Moskalenko, Lena Nabers, Sandra Terfurth and Julia Weßeler as part of their respective master's or state exam's thesis.

I would also like to thank the colleagues of the mathematics department at the Melbourne Graduate School of Education for welcoming me as a visiting researcher for three months. Thank you to my colleagues from the mathematics education department and also the faculty of mathematics at the University of Duisburg-Essen, for the many conversations: at lunch, in the hallways, while handling the media rental service or in the colloquia. The input was indispensable, especially in the development of the materials used.

Some of my colleagues from Essen deserve a special mention.

Heartfelt thanks to the entire "Arbeitsgruppe Barzel", I consider myself lucky to be part of this team.

PD Dr. Aleksandra Zimmermann has supported me since I started my teaching degree at the University nearly 10 years ago. Without the many hours of talking with her, my time at University would be not have been so good. Thank you for making my time at university more fun.

Patrick Ebers and Katharina Mros were an integral part of the lunchtime group and we also had many professional but also personal discussions outside of the lunch break. Thank you for that.

A special thanks to Hana Ruchniewicz for the countless hours of conversation—in the office, sharing a room at conferences or on social occasions. I was able to benefit from your great expertise in many fields, not only academically.

Finally I would like to thank my family, especially my parents Angela and Gerhard Göbel. They have always supported me without hesitation, even if I wanted to travel to the other side of the world for months or just needed encouragement. Thank you for everything!

Essen   Lisa Göbel
July 2020

Zusammenfassung

Technologie wird zunehmend Bestandteil des Mathematikunterrichts und der Einsatz wird auch in den Kernlehrplänen gefordert. Ein Beispiel ist die Konzeptualisierung von Parametern im Bereich quadratischer Funktionen. Die technischen Möglichkeiten lassen verschiedene Zugänge zur Manipulation der Parameter sinnvoll erscheinen (klassischer Funktionenplotter, Zugmodus, Schieberegler). In dieser Arbeit sollen diese drei technischen Visualisierungen mit einem Zugang ohne die Möglichkeit der technischen Visualisierung verglichen werden. Dazu wurde eine Guided Discovery Lernumgebung entwickelt, die in einer Interventionsstudie mit 14 Klassen der Jahrgangsstufe 9 (N = 383) durchgeführt wurde. Es zeigen sich verschiedene Stärken und Schwächen der einzelnen Visualisierungen zugunsten der dynamischen Visualisierungen durch Zugmodus und Schieberegler.

Abstract

Technology is becoming more and more integrated in mathematics teaching and the use of technology is explicitly demanded by the curricula. An example is the conceptualization of parameters in the area of quadratic functions. The technical possibilities make different approaches for manipulating the parameters seem beneficial. In this thesis, three technical visualizations (classic function plotter, drag mode, sliders) are to be compared to an approach without the possibility of technical visualization. For this purpose, a Guided Discovery environment was developed, which was tested in an intervention study with 14 classes of Year 9 (N = 383). Different affordances and constraints of the individual visualizations can be identified favouring the possibilities of dynamic visualizations through drag mode and sliders.

Contents

List of Figures

List of Tables

Part I

Introduction

Introduction

1

"(…) there is no understanding without visualization" (DUVAL 1999, p. 13)

The aim of this thesis is to investigate the interdependencies between the use of technology with its power of visualization and the ability change of representations, and students' conceptualization of parameters in quadratic functions using a discovery learning approach.

While learning mathematics objects cannot be accessed directly, but only through representations of their properties and through changing between the different representations (DUVAL 2006). Therefore, visualization, namely the identification of the relevant aspects of the representations (DUVAL 2014, p. 160), is an important part of understanding mathematics. The visualization can be supported through the use of technology for example through multiple representation systems, as they offer the possibility to display more than one representation at the same time and even provide links between the representations (see e.g. BARZEL & GREEFRATH 2015, SCHMIDT-THIEME & WEIGAND 2015).

Students nowadays live in a "digital revolution" (KMK 2016, p. 3), so the access to digital technologies can be assumed to be easier than ever before. Ninety-nine percent of German secondary school students use a smartphone or have access to one at home (FEIERABEND et al. 2018, p. 6). However, most of the school education still is with paper and pencil and scientific calculators, while computer-algebra systems, dynamic geometry systems or spreadsheets are not used regularly in Germany (SCHMIDT-THIEME & WEIGAND 2015, LORENZ et al. 2017, THURM 2020, THURM & BARZEL 2020), even though there has been a lot of research about the potentials of different aspects on using technology in the mathematics classroom (e.g PENGLASE & ARNOLD 1996, ZBIEK et al. 2007, BARZEL 2012, DRIJVERS et al. 2016, LINDMEIER 2018). Some of this research is over 20 years old (e.g. PENGLASE & ARNOLD 1996),

L. Göbel, *Technology-Assisted Guided Discovery to Support Learning*,
Essener Beiträge zur Mathematikdidaktik,
https://doi.org/10.1007/978-3-658-32637-1_1

but it is still relevant as for example graphics calculators only became mandatory in North-Rhine-Westphalia, Germany, where most of the study presented here was conducted in 2015, in 2014 and then only for the last three years of upper secondary school (MSW NRW 2012).

Digital tools can offer the possibility to dynamically change the different representations, but the question remains, if the *power of speed* that is available in this digital era is in fact beneficial for learning or if it would be better to take a step back and choose a slower option for making discoveries. With the power of speed the fast manipulation through technology, where for example students can change some aspect in some representation of the mathematical object very quickly, is meant.

Possible options to use the power of speed and available in a number of multiple representations programs are sliders and the drag mode (manipulating objects through direct dragging). Slower options are standard function plotters or even forgoing technological visualizations completely, as the students need to input function one by one or sketch them by hand. Research into the differences between the options would enable teachers and students to weigh up the potential benefits and the effort needed to use them (i.e. sliders and drag mode often have to be programmed, while function plotters just have to be turned on).

These aspects of the technological tools can be used over a number of topics. A central topic in secondary school mathematics is functions. In Germany it is one of five central topics of mathematics education in lower secondary school (KMK 2004, p. 9) and quadratic functions are the first non-linear function taught. So quadratic functions play an important role in the general understanding of functions, as they can be used as a blueprint for higher-order polynomials. Four different representations of functions are commonly used and described, namely situative-verbal, symbolic, tabular and graphical. The last three can be easily visualized and linked using multiple representation software. The parameters in quadratic functions can be manipulated in these representations e.g. through sliders (VOLLRATH & ROTH 2012) and through this manipulation the connection between parameters and transformation of quadratic functions investigated as asked for by KLINGER (2018, p. 447).

Teaching mathematics can be based on a number of approaches, one of these is discovery learning. With discovery learning students are not presented with all aspects of a concept, but asked to discover parts themselves (see Chapter 4, e.g ALFIERI et al. 2011). It is one of the methods demanded in the core curricula of North-Rhine Westphalia, Germany (MSW NRW 2007, p. 12), but there have been controversial discussions about the concept of discovery learning and its efficacy as the research offers no clear picture (e.g. MAYER 2004, ALFIERI et al. 2011). Research has, however, shown that the use of scaffolding or guidance is beneficial for the learning of mathematics (e.g. ALFIERI et al. 2011). Guided Discovery as an approach combines the openness of discovery learning with the benefits of guidance and scaffolding shown by research.

The three aspects visualization through technology, functions and Guided Discovery each play an important role and might benefit the learning of mathematics. But does a combination of these three influence the learning even more beneficial (see Chapter 5)? For the parameter concept of quadratic functions this question is investigated in the frame of this thesis.

Following this introduction (Part I) are four further main parts. The theoretical background (Part II) offers an overview over the relevant theories regarding function transformation, technology and discovery learning and current research into these fields. The theoretical background is split into the function related background (Chapter 2) regarding functions, variables and parameters as well as representations and change of representation, followed by the theoretical foundation regarding the use of technology (Chapter 3) and discovery learning (Chapter 4). Part III of this work deals with the empirical study conducted, in Chapter 5 the research questions are presented and connected to the theoretical background. Chapter 6 and 7 describe an overview of the study and the design process of the materials. Part III finishes with a description of the method of analysis (Chapter 8). The report about the results (Part IV) starts with an overview of the collected data and some information about participating classes (Chapter 9). Chapter 10, 11 and 12 concern the findings of the study. The final part of this dissertation (Part V) summarizes and discusses the results. Limitations of the study as well as implications for teaching and learning are also examined and the dissertation closes with a final conclusion (Chapter 13).

Part II
Theoretical Background

Function-related Background

2

Functions can be used to represent everyday situations, whenever the relation and dependance between two quantities are investigated and they have been described in the literature since the late 17th century (TROPFKE 1902). Conceptualizing functions can be described through different theoretical constructs (see subsection 2.1.2). One central type of function taught in secondary school are quadratic functions as the first non-linear functions taught (see section 2.2). Transformations of quadratic functions can be used as a blueprint for higher-order functions, thus understanding those transformations is an important goal to achieve (see subsection 2.2.2 and 3.4). Crucial for a full understanding of transformation are also understanding of the variable and parameter concepts (section 2.3) and being able to switch between the different representation registers (section 2.4).

2.1 Functions

Since the Meran reform from 1905, where functional thinking was established as a topic for schools in Germany for the first time (see for example KRÜGER 2002, SCHUBRING 2007, KRÜGER 2019), there has been a wide consensus that functions are a crucial topic for mathematical understanding (KRÜGER 2019). In Germany it is one of five central topics in the mathematical educational standards for both the *Mittlerer Schulabschluss*[1] and the *Allgemeine Hochschulreife*[2] passed by the Standing Conference of the Ministers of Education and Cultural Affairs of the Länder in the Federal Republic of Germany (KMK 2004, KMK 2015). Thus all students attending

---

[1] A "mittlerer Schulabschluss" can be obtained at the end of year 10 and qualifies for continuing with upper secondary education.

[2] The degree that allows one to take up any course at any tertiary institution.

© The Author(s), under exclusive license to Springer Fachmedien Wiesbaden GmbH, part of Springer Nature 2021
L. Göbel, *Technology-Assisted Guided Discovery to Support Learning*,
Essener Beiträge zur Mathematikdidaktik,
https://doi.org/10.1007/978-3-658-32637-1_2

school in Germany have to be taught the function concept. In particular students have to be able to work with functions in all representations, compare and link the different representations as well as compare different functions and use functions to solve problems (KMK 2004, p. 11 f.). These national education standards are then implemented into the mathematics curricula of the German states (called "Länder"). For the state of North-Rhine Westphalia in which most of the study presented in this thesis was conducted the state curriculum[3] for mathematics requires students to "have a basic understanding of functional dependency and use their knowledge to understand and describe relationships and changes in mathematics and the environment"[4] (MSW NRW 2007, p. 15).

Two approaches of defining functions are found in the literature (subsection 2.1.1). A number of theories describing how the learning of functions progresses have been developed. Three such theoretical approaches namely the APOS Theory (e.g. ASIALA et al. 1996) which describes the learning in a sequential way and the Grundvorstellungen (e.g. VOM HOFE & BLUM 2016) and the concept definition/image aspects (e.g. VINNER & HERSHKOWITZ 1980) are described (subsection 2.1.2). All three try to operationalize what mathematical understanding of functions means in detail.

2.1.1 Definition of Functions

Functions are usually described by one of two ways: as a one to one mapping between two sets or as covariation between two variables (LEINHARDT et al. 1990, p. 25). The now commonly used description characterizes functions as a mapping and is often called the Dirichlet-Bourbaki concept. It states that functions are "a correspondence between two nonempty sets that assigns to every element in the first set (the domain) exactly one element in the second set (the codomain)" (VINNER & DREYFUS 1989, p. 357). This description can be seen mostly in a static way as each value of the independent quantity is mapped to one value of the dependent quantity. A more dynamic view, which focusses on a change through variables, can be taken through the older definition by Bernoulli, who defined the function as such: "We call a

---

[3]The state curriculum referred to here is the one that was valid at the time of the study.

[4]German Original: "Schülerinnen und Schüler besitzen ein grundlegendes Verständnis von funktionaler Abhängigkeit und nutzen ihre Kenntnisse zum Erfassen und Beschreiben von Beziehungen und Veränderungen in Mathematik und Umwelt" (MSW NRW 2007, p. 15).

function of a variable a quantity, which is composed in any way of this variable quantity and of constants" (TROPFKE 1902, p. 143 translated by L. Göbel).[5]

2.1.2 Conceptualizing Functions

Conceptualizing the learning of functions has been tried by researchers and mathematics educators all over the world. Three approaches that have been used in a number of studies will be presented here. In mathematics education the German and English speaking communities have different traditions in conceptualizing of function i.e. the theories of *Grundvorstellungen* and *concept image/concept definition*. Both have a number of aspects in common with each other as well as with the APOS Theory. The three theories especially the Grundvorstellung one are important for the design of the study (Chapter 7).

APOS Theory is a theory, which supposes four stages: *action, process, object, schema*. The first three stages are seen as sequential, so students learning a concept would pass through all three stages, while the schema stage incorporates the three other stages. It was developed by researchers around the group of BREIDENBACH et al. (1992) (see also DUBINSKY & HAREL 1992, ASIALA et al. 1996) in order to understand how conceptualizing for example functions takes place. ASIALA et al. (1996) describe the general core of the framework as this:

> "An individual's mathematical knowledge is her or his tendency to respond to perceived mathematical problem situations by reflecting on problems and their solutions in a social context and by constructing or reconstructing mathematical actions, processes and objects and organizing these in schemas to use in dealing with the situations" (ASIALA et al. 1996, p. 7)

While developing an understanding of a new mathematical concept one acts on previously understood mathematical objects. These actions can then be internalized to processes and finally encapsulated to objects. All three components actions, processes and objects can be connected through mental schemas (ASIALA et al. 1996, p. 8). These four (actions, processes, objects, schemas) will be described now more detailed with a focus on functions.

---

[5]French Original: "On appelle Fonction d'une grandeur variable, une quantité composée de quelque manière que ce soit de cette grandeur variable et de constantes" (TROPFKE 1902, p. 143).

- *action:* When students incorporated functions on the action level, so show an action conception, they can perform transformations on mathematical objects "only by reacting to external cues that give precise details on what steps to take" (ASIALA et al. 1996, p. 9). In the case of functions students would be able to evaluate a function at specific points through substituting the input value into the formula. So the external cue would be for example the request to evaluate the function, while the reaction would be substituting the input value into the formula. BREIDENBACH et al. describe it as a static conception as each action is performed after another (BREIDENBACH et al. 1992, p. 251). The action conception is seen as an essential part of understanding and is followed by the process conception (ASIALA et al. 1996, p. 10).
- *process:* Repeating an action and reflecting on this action might lead to internalizing actions into a process. Therefore, "an internal construction is made that performs the same action, but now, not necessarily directed by external stimuli" (ASIALA et al. 1996, p. 10). Students are then even able to inverse the processes or combine a number of processes together. Having a process conception of functions means that one is able to conduct a dynamic transformation repeatedly, which will lead to the same result if started at the same point. Some parts of the process might be performed internally (ARNON et al. 2014, p. 20 f.). The process conception is important for understanding of functions as more complex functions do not give clear instructions for an action to be performed on the input (ASIALA et al. 1996, p. 10).
- *object:* Encapsulating processes into an object is the next step in understanding. Encapsulating entails reflecting on the operation, seeing the process as a whole and being able to realise "that transformations (...) can act on it and (...) actually construct such transformations" (ASIALA et al. 1996, p. 11). Students need to be able to de-encapsulate the object if necessary.
- *schema:* Organizing a number of processses and objects are then formed into a mental schema. These schemas then can be treated as objects again. In the case of functions the schema of a student would be the "totality of knowledge which for her or him is connected (conciously or subconciously)" (ASIALA et al. 1996, p. 12) to functions. So for example, it includes the methods how to compute function tables from the function equation and graphing those functions, the link between the different representations as well as characteristics of specific functions, e.g. that the graph of a quadratic function is a parabola and these parabolas can be transformed.

In the German tradition in mathematics education *Grundvorstellungen*[6] (short GV) and their development are used to describe the connection between the mathematical concepts, real world contexts and students' mental models (BLUM 2004)[7] and can be used to conceptualize the learning of functions. It was intensively described by VOM HOFE (1995) in his dissertation. VOM HOFE & BLUM (2016) characterize the concept in the following way:

"This concept describes the relationships between mathematical content and the phenomenon of individual concept formation. The numerous treatments that the GV concept has received over time nonetheless focus on three particular aspects of this phenomenon, albeit with different emphases among these treatments:

- The *constitution of meaning* of a mathematical concept by linking it back to a familiar knowledge or experiences, or back to (mentally) represented actions,
- The *generation of a corresponding mental representation* of that concept; that is, an 'internalization', which (following Piaget) enables operative action at the level of thought,
- The *ability to apply* a concept *to* real-life situations by recognizing a corresponding structure in subject-related contexts or by modelling a subject-related problem with the aid of mathematical structures."

(VOM HOFE & BLUM 2016, p. S230 emphasis in the original)

The *constitution of meaning* aspects of VOM HOFE & BLUM can be connected to the action level of APOS Theory as the meaning is intended to be developed through actions. Schemas of the APOS Theory conform with the *generation of a corresponding mental representation* of the Grundvorstellungen. Grundvorstellungen include normative, descriptive and constructive aspects. Normative aspects of Grundvorstellungen can be used to determine what a full understanding of a particular mathematical concept should include. These need to be derived through a subject-matter analysis (VOM HOFE & BLUM 2016). Descriptive aspects of Grundvorstellungen are used to describe the student's mental representation and these can include misconceptions or partial understandings concerning a specific concept (VOM HOFE & BLUM 2016). Grundvorstellungen give orientation to structure the learning process and they can be constructed through teaching, hence the con-

---

[6]Singular: Grundvorstellung, Plural: Grundvorstellungen.

[7]There are a great number of publications in German on the concept of Grundvorstellungen, for example VOM HOFE (1992), VOM HOFE (1995), VOM HOFE (2003), MALLE (2000), KLINGER (2018). If the presented aspects are also presented in the English-speaking article VOM HOFE & BLUM (2016), the English article is cited rather than the German older references.

structive aspects of Grundvorstellungen. The constructive aspects address the gap between the normative and descriptive aspects and can be characterized by the question what differences are evident and how they can be resolved (VOM HOFE 1992, p. 354).

Grundvorstellungen have been described for a number of mathematical concepts (see BLUM 2004 for a list of Grundvorstellungen of a number of concepts), in the area of functions three Grundvorstellungen have been consolidated through the last thirty years within the German community of mathematics education.

VOLLRATH (1989) already identified three characteristic aspects which can be observed while working with functions.

1. "Functions describe or create relationships between quantities: one variable is then assigned to a different one, so that one variable is considered to be dependent on the other."[8] (VOLLRATH 1989, p. 8 translated by L. Göbel)
2. "Functions capture how changes of one quantity affect a dependent quantity."[9] (VOLLRATH 1989, p. 12 translated by L. Göbel)
3. "With functions one regards a given or created relationship as a whole."[10] (VOLLRATH 1989, p. 15 translated by L. Göbel)

These three characteristic aspects have then been established as Grundvorstellungen of functional relationships by VOM HOFE (1995). BLUM (2004) named these Grundvorstellungen of a function *mapping, covariation* and *object as a whole*.[11] Mapping is aligned with a static view, where a single value of one quantity is matched to values of another quantity. Covariation, however is more aligned with a dynamic view, where a change in one quantity is observed in dependency to a change in the other quantity. Object as a whole is a global view of the function as one object (VOM HOFE & BLUM 2016), which is different to the other two ideas of mapping and covariation, as the global features of the function are considered. DOORMAN et al. (2012) identified the same three aspects of functional understanding, while

---

[8]German original: "Durch Funktionen beschreibt oder stiftet man Zusammenhänge zwischen Größen: einer Größe ist dann eine andere zugeordnet, so daß [*sic*] die eine Größe als abhängig gesehen wird von der anderen."

[9]German original: "Durch Funktionen erfaßt [*sic*] man, wie Änderungen einer Größe sich auf eine abhängige Größe auswirken."

[10]German original: "Mit Funktionen betrachtet man einen gegebenen oder erzeugten Zusammenhang als Ganzes."

[11]German original: *Zuordnung, Kovariation* and *Objekt als Ganzes*.

naming them slightly different.[12] VOM HOFE et al. (2005) describe the linking of different Grundvorstellungen as an essential requirement for developing mathematical understanding. The object Grundvorstellung can be matched to the object conception of the APOS Theory, both describe the same kind of view even though the Grundvorstellungen concept is not sequential as the APOS Theory is. So Grundvorstellungen are not necessarily developed in the order mapping, covariation and then object, while in the APOS Theory one needs to develop an action and a process conception, before developing an object conception.

A similar, but in the English speaking tradition more prominent, theoretical approach of *concept image* and *concept definition* has been developed by VINNER & HERSHKOWITZ (1980) and later TALL & VINNER (1981). As with Grundvorstellungen, concept image has a developing character and is defined by TALL & VINNER (1981) in the following way:

> "We shall use the term *concept image* to describe the total cognitive structure that is associated with the concept, which includes all the mental pictures and associated properties and processes. It is built up over the years through experiences of all kinds, changing as the individual meets new stimuli and matures." (TALL & VINNER 1981, p. 152 emphasis in the original)

So the concept image includes all mental attributes of the concept including wrong or incomplete ones. The other big part of this theory is the idea of *concept definition* which is "a form of words used to specify that concept" (TALL & VINNER 1981, p. 152). The concept image matches most of the aspects of schemas in the APOS Theory as both describe the total knowledge and connections a student has of a mathematical object.

KLINGER (2018, 2019) compared both the Grundvorstellungen and the concept image idea. For this KLINGER restricted the Grundvorstellungen to their descriptive aspects and identified several differences and similarites. Both the concept image/concept definition and Grundvorstellungen use a constructivist view on learning, but the concept image can explicitly also include wrong conceptions, while with Grundvorstellungen these are often conceived as misconceptions and not as a part of a Grundvorstellung. He concludes that the descriptive aspects of Grundvorstellungen can be seen as the fundamental core of the concept image (KLINGER 2018, p. 42; KLINGER 2019, p. 71).

---

[12]They name the three aspects mapping, covariation and object as a whole "input-output-assignment", "dynamic process of co-variation", "mathematical object" respectively (DOORMAN et al. 2012, p. 1246).

2.2   Quadratic Functions

Quadratic functions are the first non-linear function taught in secondary school (KMK 2004), therefore they are the first function with a local minimum (or maximum if the lead coefficient is negative) and a non-constant rate of change. As a new object, they allow new tasks for example in modelling of non-linear relationships, e.g. calculating the best distance and angle for a penalty shot in basketball, but can also be used for understanding geometric concepts such as conic sections and focal points. Thus quadratic functions offer rich and manifold activities and connections.

2.2.1   Definition of Quadratic Functions

A quadratic function is a polynomial of degree two most often written in the standard form (see below). Graphs of quadratic functions are called parabolas. The easiest quadratic function is $f(x) = x^2$, the graph of this function is often called the standard parabola (WITTMANN 2008).

Standard- and Vertex Form
Quadratic functions can be expressed symbolically in different forms. Two common forms are the standard- and vertex form. The standard form is $f(x) = a \cdot x^2 + b \cdot x + c$, where a, b, c, with $a \neq 0$ are arbitrary real numbers and the vertex form is often written as $f(x) = a \cdot (x - h)^2 + k$ with $a, h, k, a \neq 0$ arbitrary numbers. The latter form is named vertex form as the vertex of the parabola can directly be determined as (h, k). In this thesis, we will name the parameter in the vertex form a, b, c respectively so $f(x) = a \cdot (x - b)^2 + c$, where (b, c) is the vertex of the parabola.[13] Most students recognize the advantage of the vertex form over the standard form, while trying to connect the graphical to algebraic representations (DREYFUS & EISENBERG 1987), most likely as they can identify the vertex point directly from it. The different algebraic forms of quadratic functions are suited for different purposes. While the standard form offers potentials for innermathematical concepts as blueprints for any higher order polynomial functions regarding for example the y-intercepts, the vertex form and transformation of quadratic functions in vertex form can be used as a prototype for transformation of any function.

---

[13]The denotation of standard form and vertex form as well as the naming of parameters in the vertex form is kept in the following. If the presented literature differs in naming it is adapted to the here used vocabulary to avoid confusion.

The core curriculum of North-Rhine Westfalia, one of 16 states in Germany and the state in which all but one of the schools participating in this study are situated, includes a number of competencies regarding quadratic functions (MSW NRW 2007). These competencies clarify the learning goals stated by the educational standards on a federal level (KMK 2004). The competencies for quadratic functions are supposed to be achieved at the end of year 9 and will be stated in the following. Students should be able to

- "describe (...) quadratic functions in their own words, with function tables, graphs and terms, changing between these representations and naming the advantages and disadvantages of these representations"[14] (MSW NRW 2007, p. 31 translated by L. Göbel)
- "Students interpret the parameters of the algebraic representations of (...) quadratic functions in the graphical representation and use this in applied situations"[15] (MSW NRW 2007, p. 31 translated by L. Göbel)
- "use (...) quadratic functions to solve outer- and innermathematical problems"[16] (MSW NRW 2007, p. 31 translated by L. Göbel).

ZASLAVSKY (1997) identified ten aspects through analysis of mathematics curricula, textbooks and reports of teachers, which are important for conceptualization of quadratic functions:

"Common algebraic forms of a quadratic function; Connections between *x-intercepts* of a parabola and the coefficients of its algebraic form; Conditions determining the location of the *x-intercepts* of a parabola; Conditions determining the number of *x-intercepts*; Conditions determining the location of the *y-intercept* of a parabola; Conditions determining the type of concavity of a parabola; Symmetrical properties of a parabola; Extreme values of a quadratic function; Connections to a linear function; Special cases of pairs of quadratic functions" (ZASLAVSKY 1997, p. 22 emphasis in the original)

The ten aspects will be explained in more detail below.

---

[14]German original: "Schülerinnen und Schüler stellen (...) quadratische Funktionen mit eigenen Worten, in Wertetabellen, Grafen und in Termen dar, wechseln zwischen diesen Darstellungen und benennen ihre Vor- und Nachteile."

[15]German original: "Schülerinnen und Schüler deuten die Parameter der Termdarstellungen von (...) quadratischen Funktionen in der grafischen Darstellung und nutzen dies in Anwendungssituationen."

[16]German original: "Schülerinnen und Schüler wenden (...) quadratische Funktionen zur Lösung außer- und innermathematischer Problemstellungen an."

1. *Common algebraic forms of a quadratic function:*
   There are different algebraic representations of a quadratic function, all of which
   have different potentials for the learning. The three common ones identified by
   ZASLAVSKY (1997) are the standard and vertex form described above and a
   factored form $f(x) = a \cdot (x - d) \cdot (x - e)$, where d and e are the x-intercepts
   (p. 39). For the study presented in this thesis the vertex form will be used.

2. *Connections between x-intercepts of a parabola and the coefficients of its algebraic form:*
   For a given quadratic function in the standard form the x-intercepts can be determined through inputting the parameters of the algebraic representation in solving
   formulas (ZASLAVSKY 1997, p. 39). These formulas can be used to illustrate the
   connection between the x-intercept and the parameters. A common solving formula taught in Germany is the pq-formula (as defined in e.g GERSEMEHL et al.
   2013, p. 28), where for a function of the form $f(x) = x^2 + p \cdot x + q$ the two
   x-intercepts can be determined through $x_{1/2} = -\frac{p}{2} \pm \sqrt{\left(\frac{p}{2}\right)^2 - q}$.

3. *Conditions determining the location of the x-intercepts of a parabola:*
   Quadratic functions can have, no, one or two x-intercepts. In cases of two intercepts the x-intercepts can be on the same side of the y-axis, one can be on the
   y-axis or they can be on different sides of the y-axis (ZASLAVSKY 1997, p. 40).

4. *Conditions determining the number of x-intercepts:*
   How many x-intercepts the quadratic function has, can be determined through
   the discriminant, which can be computed for the function $f(x) = a \cdot x^2 + b \cdot x + c$
   with $\Delta = b^2 - 4a \cdot c$. If the discriminant is positive, the function has two x-
   intercepts, if it is negative, the function has no real x-intercepts and if it is zero,
   the vertex point is on the x-axis (ZASLAVSKY 1997, p. 40). Knowledge about the
   number of intercepts can for example help while sketching the function.

5. *Conditions determining the location of the y-intercept of a parabola:*
   Not all parabolas have x-intercepts, but for all of them a y-intercept can be
   computed. The location of the y-intercept depends on the value of parameter c
   in the standard form, so the y-intercept is above the x-axis for $c > 0$, on the
   x-axis for $c = 0$ and below the x-axis for $c < 0$ (ZASLAVSKY 1997, p. 41).

6. *Conditions determining the type of concavity of a parabola:*
   Depending on the value of the lead coeffficent a in all three common algebraic
   forms, the parabola opens upwards or downwards. For $a > 0$ the parabola is
   concave-up, while it is concave-down for $a < 0$ (ZASLAVSKY 1997, p. 41). The
   direction of concavity is one of the influences of parameter a that is investigated
   in the study presented in this thesis.

7. *Symmetrical properties of a parabola:*
   All parabolas are symmetric with a symmetry axis parallel to the y-axis. If the vertex point is on the y-axis, the y-axis is also the symmetry axis. Identifying the symmetry axis is beneficial for the determining of x-intercepts as the two x-intercepts are symmetrical to the symmetry axis, so if one is determined the other can be deduced using the symmetry (ZASLAVSKY 1997, p. 42). The symmetry can also help, when asked to sketch the graph. If one knows the symmetry axis it suffices to compute values on one side of it and the other side of the parabola can again be deduced through symmetry.

8. *Extreme values of a quadratic function:* Every parabola has an extreme value, so either a maximum or a minimum value depending on the sign of the parameter a in all three common algebraic forms. It is the vertex point and if a is negative it is a maximum, while it is a minimum for a positive a (ZASLAVSKY 1997, p. 43).

9. *Connections to a linear function:*
   One difference between linear and quadratic functions is that there can be no three colinear points, so three points which could all be on the same linear function, on a parabola (ZASLAVSKY 1997, p. 43). As quadratic functions are often the first non-linear functions taught, this is an important point in the conceptualization of quadratic functions, otherwise this might lead to misconceptions and over-generalizing (see subsection 2.2.3).

10. *Special cases of pairs of quadratic functions:*
    Recognizing the connections between different parabolas is important for the conceptualization of the parameter concept. So for example, if the two parabolas only differ in parameter c in the standard form, they have the same symmetry axis and the same concavity. One can be produced through a vertical translation of the other (ZASLAVSKY 1997, p. 43).

The ten aspects by ZASLAVSKY (1997) operationalize the quadratic functions concept more concrete than the competencies in the core curricula, giving concrete aspects that are important for the learning of quadratic functions. The core curriculum takes a broader view and focusses more on the linking of representations and the use of the functions in applied situations. But to be able to use quadratic functions in these situations and solve inner- and outermathematical problems, the students need to take the ten aspects into account. For this thesis, all aspects are relevant to some degree. As students have only encountered the standard parabola before the intervention, the discriminant and solving formulas most likely won't be introduced yet. For the intervention the aspects 5, 6 and 10 are especially important, as they are directly related to the topic of influence through parameters on quadratic functions in the vertex form.

But how does conceptualizing quadratic functions and transformations of those functions occur? This is elaborated in the next subsection.

2.2.2   Conceptualizing Quadratic Functions with Focus on Transformation

Every quadratic function can be obtained by transforming the quadratic function $f(x) = x^2$. Even though extensive research has been conducted on the concept of functions itself, research on transformations of quadratic functions before the year 2000 was sparse (BAKER et al. 2000).

EISENBERG & DREYFUS (1994) investigated, how students visualized function transformations. Students in the study had condsiderable difficulties transforming functions. However, vertical transformations seem to be easier than horizontal transformations. EISENBERG & DREYFUS (1994) explain this with the "complexity of the statements themselves, for much more is involved in visually processing the transformation of f to $f(x + k)$ than in visually processing the transformation to $f(x) + k$" (EISENBERG & DREYFUS 1994, p. 58). As well as horizontal transformations being more difficult than vertical transformation, most students in the study were not able to transform other functions than quadratic functions. The authors posit that to understand transformations the students need to consider functions as objects and not as processes as described in subsection 2.1.2, which might not be the case if the parent function is more complicated (EISENBERG & DREYFUS 1994). Extending the APOS Theory to transformations they identify an incomplete process conception of transformations. One reason for this might be that students did not see the transformation dynamically so one graph transforming into another one, but more as a static product (EISENBERG & DREYFUS 1994).

Also using the APOS Theory as a theoretical construct BAKER, HEMENWAY, & TRIGUEROS (2000, 2001) adapted it for the study of transformation. Understanding the concept on an *action level*, "students are able to perform transformations on functions in an analytical context by substituting values in them one by one and by drawing the graph of the function based on the evaluation of independent points" (BAKER et al. 2000, p. 42). However, the students need to have the visualization in front of them to be able to determine the important properties of the transformed function. The authors conjecture that they have then a more static view on transformation (BAKER et al. 2001). When students reach the *process level*, they are able to evaluate functions and changing the properties of simple functions through transformations. They have a more dynamic view including intermediate steps of transformations of basic functions (BAKER et al. 2001). Finally, understanding the concept on an *object level*, students can predict the outcome of transforming any functions (BAKER et al. 2000, p. 43).

BAKER et al. (2000) then use the APOS Theory to analyse, how the use of graphing calculators in connection with writing tasks influence university students' learning of transformation. The results of their study show that only few students understood transformations on an object level, as well as supporting the results of EISENBERG & DREYFUS (1994) that transformations of linear or quadratic functions are easier than more complex transformations and vertical transformations are easier than horizontal. A possible reason based on the APOS Theory for the various difficulties of vertical and horizontal transformations is also given by the authors:

> "vertical transformations are actions performed directly on the basic functions while horizontal transformations consist of actions that are performed on the independent variable and a further action is needed on the object resulting from the first action to get the result of the transformation" (BAKER et al. 2000, p. 46).

Applying the APOS Theory with focus on the process- and object parts on to quadratic functions, METCALF (2007) describes, how graphing the graph of $f(x) = x^2$ starts on a process level through computing ordered pairs from the domain and then graphing them. Once it is encapsulated as an object, the graph can be manipulated. Students can then transform the function $f(x) = x^2$ to more difficult quadratic functions e.g. $f(x) = -(x-2)^2 + 1$ on a process level. After this is again encapsulated on an object level and for example use the graph of $f(x) = -(x-2)^2 + 1$ to solve a problem, students can move on to higher order polynomials (METCALF 2007, p. 41).

Using the different approach of flexibility to describe the complexity of transformations, KIMANI (2008) investigated high school and college students' understanding of function transformation, function composition, function inverse, and their connections. Regarding transformations he found five aspects of students understanding:

- Students showed no conceptual understanding, but only used formulas and relied on rules taught to them, which leads to problems when trying to generalize the results (KIMANI 2008, p. 231 f.).
- Students preferred to think structurally, this means that the students compared the structure of graphs in order to determine transformations, thus preferring the graphical representations (KIMANI 2008, p. 234 f.). They treated graphs as complete objects, which is one of the three Grundvorstellungen of functions described by VOM HOFE & BLUM (2016) as well as the object conception of the APOS Theory.

– Scaling transformations, so vertical or horizontal stretching of graphs were the
  most difficult kinds of transformations (for example a vertical scaling could be
  $f(x) = a \cdot x^2$). Students were able to identify them, but could not write an
  equation and the horizontal stretches were more difficult than the vertical. The
  operational nature and the need to invoke proportional thinking are posed as
  possible reasons for the difficulties (KIMANI 2008, p. 236 f.). KIMANI's findings
  substantiate the findings of EISENBERG & DREYFUS (1994) and BAKER et al.
  (2000) that horizontal transformations are more difficult than vertical transformations.
– Students "showed facility with multiple representations in their thinking" (p. 231)
  and were able to change between representations. As students preferred to think
  structurally, the students changed from function equation to the graphical representation (KIMANI 2008, p. 238 f.).
– The study also showed that an incomplete understanding of the function concept
  impacts students' understanding of functions transformations (KIMANI 2008,
  p. 239 f.).

Trying to give insight not only into the learning processes and explanations of
students, but also investigating the explanations of teachers, ZAZKIS et al. (2003)
presented the teachers and students the two functions $y = x^2$ and $y = (x - 3)^2$. A
common mistake when asked how the second function is positioned compared to
the first, is that it is shifted three units to the left, whereas in reality it is shifted to the
right. The researchers analysed common trends in explanations, attitudes towards the
difference between intuition and reality, and possible discrepancies in the first two
depending on the experience through comparing pre-service, in-service teachers,
and high school students (ZAZKIS et al. 2003). While most students did not explain
the counterintuitive transformation, the teachers' responses could be classified into
four categories: *citing rules, point-wise approach, attending to zero and "making
up", transforming the axes* (ZAZKIS et al. 2003, p. 442 f.). Answers falling into the
citing rules category consisted of only stating the rule that the graph of $y = (x - 3)^2$
is moved three units to the right of the graph $y = x^2$, but otherwise looks the same.
The citing rules category corroborates the results of KIMANI (2008) regarding that
students relied on rules. If the teachers are only citing rules as explanations it is not
surprising that students relied on the rules to understand transformations.

Creating the graph of the function using a function table and then plotting the
values of this table was classfied as a point-wise approach and the teachers in the
study saw potential of this approach as an explanatory tool (ZAZKIS et al. 2003).
Using not a complete function table, but rather the x-intercept, i.e. the vertex and then
arguing with symmetry and shape of a parabola, was classified as attending to zero

and then "making up". For an explanation of this approach it was then described what happens to all other points of the graph. The last category of explanations was assigned if teachers did not describe moving the graph to the right, but rather moving the y-axis to the left and therefore seemingly move the graph to the right. This explanation only occurred in a very small number of cases. ZAZKIS et al. (2003) also identified the difference that in-service teachers were able to provide the explanations more promptly and also had a bigger variety of explanations than pre-service teachers. This is understandable as in-service teachers most likely had encountered more difficulties in teaching the horizontal transformation and had to adapt their explanations to foster understanding (ZAZKIS et al. 2003).

2.2.3   Obstacles and Misconceptions

There have been a number of studies identifying misconceptions and obstacles in the field of quadratic functions. In an Israeli large-scale study ZASLAVSKY (1997) identified the following five main conceptual obstacles concerning quadratic functions:

1. "The interpretation of graphical information (…);
2. The relation between a quadratic function and a quadratic equation.
3. The analogy between a quadratic function and a linear function.
4. The seeming change in the algebraic form of a quadratic function whose parameter is zero.
5. The over-emphasis on only one coordinate of special points." (ZASLAVSKY 1997, p. 29).

Along this classification the relevant misconceptions and obstacles will be presented.

The Interpretation of Graphical Information
GOLDENBERG (1988) described a number of illusions and obstacles appearing while using graphical representations of functions. Students working with transformed parabolas used the notion of height to describe the vertical position of a parabola in the coordinate plane. However, due to confusion between the notion of height and the y-intercept in the standard form, the students are led into a wrong direction. GOLDENBERG (1988) explains this with a not robust enough understanding of the transformation to overcome the difficult perception and recommends a sensible use of language and in this case the notion of height in coordinate planes. Another source of illusions are graphs without certain points on them and therefore a vertical translation of a linear function might be mistaken for a horizontal. In the case of

parabolas, scale of the window is also important, as this influences the perception of the shape of the graph. One example of this are two graphs of two quadratic functions which only differ in parameter c (e.g. $f(x) = x^2$ and $f(x) = x^2 + 1$). If the two graphs are viewed at the same time in the same window it appears that the upper parabola is blunter (GOLDENBERG 1988, p. 148). He proposes that if the two graphs are constructed through manipulation of the graph for example through dragging at the graph this illusion might vanish. ELLIS & GRINSTEAD (2008) also encounter this problem, which seems to be related to the chosen window in most graphical programs, while PINKERNELL (2015, see also PINKERNELL & VOGEL 2016, 2017) elaborates further on this misconception regarding parameter c in $f(x) = x^2 + c$. The interviewed students argue that the graphs are moved upwards as well as changed in width (PINKERNELL 2015, p. 2532).

Conforming with research results in other areas of functions (e.g. graph as a picture mistake as identified for example by JANVIER 1978), students might also interpret the graph of a quadratic function in the manner that the parabola ends at the end of the visible window, so they might assume it does not have a y-intercept if none is shown (ZASLAVSKY 1997, p. 30 f.) This obstacle relates to the source of illusions described by GOLDENBERG (1988), which is also an interpretation of the graphical information on the screen.

The Relation Between a Quadratic Function and a Quadratic Equation
The equations $2x^2 + 4x - 6 = 0$ and $x^2 + 2x - 3 = 0$ have the same solution, however the functions $f(x) = 2x^2 + 4x - 6$ and $f(x) = x^2 + 2x - 3$ are not the same functions. This discrepancy between knowledge learnt in the field of quadratic equations and knowledge learnt in the field of quadratic functions presents a conceptual obstacle (ZASLAVSKY 1997, p. 31 f.).

The Analogy Between a Quadratic Function and a Linear Function
If linear functions are taught in the form $y = a \cdot x + b$ and then the quadratic function in the form $y = a \cdot x^2 + b \cdot x + c$, students can be confused due to the use of the same letter as parameter with different properties. This might lead to calculation of a in quadratic functions incorrectly with the rise over run[17] method (ZASLAVSKY 1997, p. 32 f.). However, over-emphasizing the analogy between the linear and quadratic functions might also lead to parameter stereotypes, especially if the parameters are always named the same. If these stereotypes are prominent, so for example c as the

---

[17]The rise over run method is used to calculate the slope of linear functions with two given points of this function. So, for example, if the points $(2, 5)$ and $(4, 8)$ are given, the slope m can be calculated with $m = \frac{8-5}{4-2} = \frac{3}{2}$.

y-intercept for linear functions, it can lead to discomfort and even problems in the solving process, if the letter c is to be used in a different context (BILLS 1997).

ELLIS & GRINSTEAD (2008) also identify one misconception which is based on ZASLAVSKY's (1997) third obstacle regarding the analogy between linear and quadratic functions. They investigated students' generalizations about connections between algebraical and graphical representations of quadratic functions in the standard form ($f(x) = a \cdot x^2 + b \cdot x + c$). Special focus was on the role of the parameters a, b, c. Findings show that two-thirds of students that participated in semi-structured interviews after about two weeks of instruction identified a as the slope of the parabola in analogy to linear functions and, thus, showing parameter stereotyping. Other interpretations of the role of a were that a influences only the shape of the graph, which is incorrect for the standard form, but would be correct for the vertex form, a is a stretch factor again without realising that a influences the vertex, a influences the opening of the parabola. These three interpretations were linked to sequences in the teaching. The interpretation of a as the slope however was not connected to the instruction, one possible reason is overgeneralizing from linear functions. ELLIS & GRINSTEAD's (2008) analysis identified three aspects in the instructional unit, which could lead to this misconception: "use of linear analogies, the rise over run method, and the dynamic view" (ELLIS & GRINSTEAD 2008, p. 290).

For parameters b and c the students gave different interpretations of the role, some (partly) correct, some incorrect. For parameter b students interpreted it as affecting the vertex, translating it horizontally, as the width of the parabola, the x-intercept or as the slope of the graph. ELLIS & GRINSTEAD (2008) trace the interpretation of b as the slope back to overgeneralizing from the linear functions, which these students were taught in the form $y = a + b \cdot x$. Parameter c is interpreted either correctly as the y-intercept, or incorrectly as the x-intercept, as translating the graph or as affecting the vertex/being part of the vertex. The last two interpretations can be attributed to the vertex form (ELLIS & GRINSTEAD 2008).

The Seeming Change in the Algebraic Form of a Quadratic Function Whose Parameter is Zero.
If one of the parameters b or c in $f(x) = a \cdot x^2 + b \cdot x + c$ is zero and therefore not written in the equation, students might mistake those functions as not quadratic, even though they are special cases (ZASLAVSKY 1997).

The Over-emphasis on Only One Coodinate of Special Points
When asked if two parabolas share the same vertex, students base their answer mistakenly only on the x-coordinate of the vertex, so vertical translation are ignored.

One reason for this obstacle might be too much emphasis on the relevant nonzero coordinate of the x- and y-intercepts (ZASLAVSKY 1997).

Relevance for This Thesis
For this thesis the obstacles regarding the interpretation of graphical information and the analogy between linear and quadratic functions are most relevant, as these are directly linked to the intention of the intervention. The use of analogies helps in the conceptualization of the new concepts, but it also could result in misconceptions and thus problems and obstacles. If students interpret the graphical information wrongly, it might even lead to wrong deductions about the influence.

More obstacles and risks that are related to using technology not only in the frame of quadratic functions will be presented in more detail in Chapter 3.

2.3   Variables and Parameters

Variables and parameters have been described in many studies over the years (see for example KÜCHEMANN 1981, WAGNER 1983, BLOEDY-VINNER 2001). It becomes evident that the underlying concepts are very difficult for learners. The reasons for the difficulties can be found in the manifold ways in which variables and parameters are used when learning and doing mathematics. Therefore, the concept of variables has been described by a number of different roles and aspects. While variables can take different roles, in these different roles, various things can be done with the variables, these activities are called aspects.[18] In this section we will present the different approaches which describe the concept of variables. Then, more specifically, the focus will be on parameters as these can be seen as an example how the different aspects and roles of variables come into play, before describing the connection between the two concepts.

2.3.1   The Concept of Variables

Variables have been defined in a number of different ways. SCHOENFELD & ARCAVI (1988) list ten different definitions of variables from the years 1710 to 1986[19],

---

[18]GREEFRATH et al. (2016) however, define aspects as parts of a concept through which it can be technically described (p. 17).

[19]The definition collected by SCHOENFELD & ARCAVI (1988) from 1986 originates not from mathematics but from computer sciences and thus the second newest from 1984 was chosen for this thesis.

which show the spectrum of the concept. A very early definition of variables is the following:

> "Variable quantities ... are such as are supposed to be continually increasing or decreasing: and so do by the motion of their said increase or decrease, generate lines, areas or solidities." (HARRIS 1710 as cited in SCHOENFELD & ARCAVI 1988, p. 421)

In this definition the variable can be seen as being only used in the literal sense of variable, so as a changing quantity.

Whereas in 1984 the entry for variable in the *Dictionary of Mathematics* states the following definition:

> "A general purpose term in mathematics for an entity which can take various values in any particular context. The domain of the variable may be limited to a particular set of numbers or algebraic entities. In the same context a constant is a quantity which is restricted to a single value in any one expression. Thus in the real equation $ax + b = 0$, a and b are given as constants and x, the variable, is any number from the set $\mathbb{R}$ which in the context makes the statement true. If one variable is expressed in terms of another, as in $y = x^2$, the one whose values may be assigned at will is the independent variable, the other is dependent." (GLENN & LITTLER 1984, p. 227)

The definition by GLENN & LITTLER (1984) focuses in the first two sentences more on variables in a generalised number sense, however, in the example of equation the variable x is treated more in an unknown sense and the parameters a and b are viewed as constants. These foci already incorporate the different roles variables can play and will be explained below. SCHOENFELD & ARCAVI (1988) outline that some definitions, e.g. the one given by Harris in 1710, try to "give the reader a sense of 'how it works'" (p. 423). This also applies to the later part of the definition by GLENN & LITTLER (1984), but theirs also leads to new technical terms that need to be discussed (here "domain of the variable" or independent/dependent variable). In general, using other technical terms that need clarification in addition to the one which is the subject of the definition can lead to difficulties. In these cases, it is questionable if one can be understood without the other (SCHOENFELD & ARCAVI 1988, p. 423). This makes defining variables clearly difficult to impossible.

HART (1951 as cited in USISKIN 1988) defines variables more in the style of GLENN & LITTLER (1984) as a "number that may have two or more values during a particular discussion" (p. 8). As with nearly every definiton, this one covers just one but not all features of variables. A newer definition can be found in SPECHT (2009):

> "As a working definition, in the following a variable is understood to mean a general object that can take on values (for example numbers, fields, and algebras over a field). It can be represented by words, letters or other symbols. In different contexts, different

conceptions of variables and variable aspects come into play. Variable is understood as a generic term for certain unknown, indeterminate and changing quantity. Variables can be viewed under the situational, the substitution and the calculation aspect." (SPECHT 2009, p. 39, translated by L. Göbel)[20]

In her definition, SPECHT (2009) includes the three often described roles of variable, variable as certain unknowns, indeterminate (often used as a synonym for generalizer) and changing quantity as well as in the German mathematics education community established variable aspects as described by MALLE (1993). The definition corresponds to the type that gives readers a sense of how one works with variables (see also SCHOENFELD & ARCAVI 1988). While solving an equation, the variable takes on a different role compared to when a general situation is described. In the equation the variable is the unknown and the value is sought after. In contrast, when a general situation is described usually more than one set of values satisfies the situation. Because variables can be used in a number of different ways depending on the context, a rigorous definition without using other concepts or describing how to use variables seems to be impossible. The variable can change its meaning depending on the situation it is used in (HECK 2001). Of the variety of different aspects, three different roles of a variable, namely variable as generalizers, unknowns or changing quantity have been widely described (FREUDENTHAL 1983, KIERAN 1992, USISKIN 1988, DRIJVERS 2001, HECK 2001). However, different authors describe more than these three roles. A selection of these shall be presented in more detail as all roles are relevant to understand the complexity of the concept of variables.

KÜCHEMANN (1981) identified in a study with 13- to 15-year-old students six ways in which the children use letters in mathematical tasks. They interpret letters in the following ways:

– *Letter evaluated*, where a numerical value is assigned, e.g.: "What can you say about a if $a + 5 = 8$?" (KÜCHEMANN 1981, p. 105), in this task the students can give the unknown a value, without operating on the unknown.
– *Letter not used*, where it is ignored, e.g. in the task: "If $a+b = 43$, $a+b+2 = \ldots$" (KÜCHEMANN 1981, p. 106), the variables can be ignored through focussing on

---

[20]German original: "Als Arbeitsdefinition wird im Folgenden unter einer Variablen ein allgemeines Objekt verstanden, das Werte annehmen kann (beispielsweise Zahlen, Körper und Algebren). Sie kann durch Worte, Buchstaben oder andere Symbole repräsentiert werden. In unterschiedlichen Kontexten kommen verschiedene Variablenauffassungen und Variablenaspekte zum Tragen. Variable wird dabei als Oberbegriff für bestimmte Unbekannte, für Unbestimmte und für Veränderliche verstanden. Variablen können unter dem Gegenstands-, dem Einsetzungs- und dem Kalkülaspekt betrachtet werden." (SPECHT 2009, p. 39)

the $+2$ operation and then only taking the right hand side of the first equation into account (KÜCHEMANN 1981, p. 106).

– *Letter used as an object*, as a shorthand for an object, e.g. a for apple and b for banana in "$2a + 5b + a = 3a + 5b$" (KÜCHEMANN 1981, p. 107).

– *Letter used as a specific unknown*, which can be directly operated upon, e.g. in the task "add 4 onto $n + 5$" (KÜCHEMANN 1981, p. 108).

– *Letter used as a generalised number*, where it represents several values, e.g. in the task "What can you say about c if $c + d = 10$ and c is less than d" (KÜCHEMANN 1981, p. 109), where students could give systematic lists of possible values or even a general term $c = 10 - d$ (KÜCHEMANN 1981, p. 109).

– *Letter used as a variable*, where it represents a range of values and "a systematic relationship" (KÜCHEMANN 1981, p. 104) exists, e.g. in the task "Which is larger, $2n$ or $n + 2$?" (KÜCHEMANN 1981, p. 111). This view is a more dynamic one and connects all possible values, if the letter is treated as a generalised number as well as establishing a relationship between those values.

In contrast to KÜCHEMANN (1981), FREUDENTHAL (1983) only distinguishes the uses of variables between *polyvalent names, placeholders* and *variable objects*. While polyvalent names are used to describe general statements that apply to all objects that are represented, variable objects actually represent something that varies (FREUDENTHAL 1983, p. 491 f.). In some works he differs at polyvalent names between unknown and indeterminate, where unknowns are used in equations, while indeterminates are used for general propositions (FREUDENTHAL 1962, FREUDENTHAL 1973, p. 262).

USISKIN (1988) describes some of the same roles as KÜCHEMANN (1981) with different terms, but similar meaning. USISKIN (1988) distinguishes variables as pattern generalizers, unknowns, parameters and arbitrary marks on paper (USISKIN 1988). Küchemann's letter evaluated and Usiskin's unknown both describe the same use of a variable, namely the substitution of a value and Küchemann's letter as a specific unknown describes the same use as Usiskin's arbitrary marks on paper. Freudenthal's indeterminate matches most aspects of Küchemann's generalised number and Usiskin's pattern generalizers.

MALLE (1993) describes not only roles of, but also aspects of variables, which can be related to the roles. He differs between five different aspects, which will be described below. The original German name of the aspects are stated behind the translation (MALLE 1993, p. 46 f.).

– *situational aspect* (Gegenstandsaspekt)
A variable is seen as an unknown or indeterminate number. Viewing a term e.g. $\frac{x+y}{2}$ in regard to this aspect results in an indeterminate number, as it is only

known to be the arithmetic mean of two arbitrary numbers. An equation is seen
as a statement regarding known or unknown numbers (MALLE 1993, p. 46 f.).
- *substitution aspect* (Einsetzungsaspekt)
The variable is seen as a placeholder for numbers, in which the numbers can be
substituted. For example, the term $\frac{x+y}{2}$ is viewed as a kind of number, which
is transformed into a number, once values are substituted for x and y, while an
equation, e.g. $x + y = 8$ is viewed as a kind of statement, which is transformed
into a true or false statement, once values are substituted for the variables (MALLE
1993, p. 46 f.).
- *calculation aspect* (Kalkülaspekt)
The variable is a meaningless symbol, which can be operated on following certain rules. Both terms and equations are viewed as string of characters, whose
significance can be ignored while operating on them within the limits of certain
rules (MALLE 1993, p. 46 f.).
- *single-value aspect* (Einzelzahlaspekt)
With this aspect variables are applied on to a fixed set. The variable is an arbitrary,
but fixed number and represents one value only. It can be chosen randomly out
of the set, but it is fixed thereafter (MALLE 1993, p. 80.).
- *area aspect* (Bereichsaspekt)
The variable is seen as an arbitrary number from the set, while representing every
number of the set. The area aspect is split into two subaspects
  - *simultaneous aspect* (Simultanaspekt)
  All values from the predetermined set are represented at the same time.
  - *changing-quantity aspect* (Veränderlichenaspekt)
  All values from the predetermined set are represented in chronological order,
  but the set is run through in a given way.
While the simultaneous aspect is seen when a variable is used in combination
for example with the universal quantifier ($\forall x \in \mathbb{R}$), the variable aspect is used
in functions (MALLE 1993, p. 80 f.). According to MALLE the changing-quantity
aspect is indistinguishable from functional thinking (MALLE 1993).

These aspects of variables are a widely used conceptualization of variables in the
German mathematics education community. They help to clarify the ways one
can operate with variables and how important it is to interpret the variable in the
given contexts (see also BARZEL & HOLZÄPFEL 2017). MALLE (1993) and SIEBEL
(2005) relate the aspects described above unambiguously to the roles of variables
described by KÜCHEMANN (1981), but SIEBEL (2005) assigns each of the aspects as
described by Malle exactly one role described by KÜCHEMANN (1981), this contra-
dicts MALLE's (1993) description where both the *letter evaluated* and *letter used*

as a specific unknown are assigned to the single-value aspect. Other authors, for example BARZEL & HOLZÄPFEL (2017) claim that for the situational, substitution and calculation aspect, variables can be used in more than one role of variables (placeholder, changing quantity, generalised number). They connect the ideas of aspects and roles of variables by describing what roles the variables can take and what can be done with it. So for example for the variable as a generalizer regarding the substitution aspect, arbitrary numbers can be substituted (BARZEL & HOLZÄPFEL 2017, p. 4).

URSINI & TRIGUEROS (2001) developed a framework for the understanding of variables. In this "three uses of variable", short 3UV-model they operationalize what understanding the concept of variable means. They differ between variable as unknown, as a general number and in functional relationships and list for each of the three uses the requirements they find necessary to understand the concepts (URSINI & TRIGUEROS 2001, URSINI & TRIGUEROS 2004). For example for understanding the variable as unknown, students need to "recognise and identify in a problem situation the presence of something unknown that can be determined by considering the restrictions of the problem" (URSINI & TRIGUEROS 2001, p. 328). For a full list of all requirements for all three uses see URSINI & TRIGUEROS (2001).

2.3.2 The Concept of Parameters

As with variables, parameters have also been defined in many ways. Some authors describe parameters as a role of variable (e.g. USISKIN 1988) others describe it as a higher order variable with different roles itself (BLOEDY-VINNER 2001, HECK 2001, VAN DE GIESSEN 2002, DRIJVERS 2003).

The parameter concept can also be described as an "elusive concept that carries with it the difficulties of literal symbols and the ambiguities of its analogy-difference with the concepts of variable and unknown" (FURINGHETTI & PAOLA 1994, p. 368).

The entry in the Dictionary of Mathematics defines parameters as

> "a variable to which other variables are related. This can be used either to obtain the other variables if required, as in parametric equations, or, in statistical parameters such as mean or standard deviation, to describe a set of variables collectively. (...)" (GLENN & LITTLER 1984, p. 147 f.)

Others define parameter as something that "stands for a number on which other numbers depend" (USISKIN 1988, p. 14), or as "a constant that may change" (VAN DE GIESSEN 2002, p. 97). While USISKIN (1988) classifies parameters as a variable role, VAN DE GIESSEN (2002) and DRIJVERS (2001) describe it as something with different roles itself. DRIJVERS (2003) even speaks of parameters as meta-variables.

There are a number of difficulties which can arise during the conceptualization of parameters. Students often struggle with the paradoxical epistemic nature of a parameter in the way that "it is a fixed, particular number, yet it remains indeterminate in that it is not an actual number" (BARDINI et al. 2005, p. 130).

Two difficulties while understanding the concept of parameter are that the role of parameters and variables are context related and depending on the situation change during the problem solving-process (BLOEDY-VINNER 2001). For example in the task "In the following equation x is an unknown and m is a parameter: $m(x - 5) = m + 2x$. For what value of the parameter m will the equation have no solution." (BLOEDY-VINNER 2001, p. 178) the parameter and variable both take on briefly the role of the unknown while solving the task. Parameters can also not be explained without second-order functions or quantifiers, which presents difficulties due to the logical complexity (BLOEDY-VINNER 2001, p. 182). In the case of the parameters inducing a function of the second order (DRIJVERS 2003), the hierarchical position of parameter is higher than of an ordinary variable.

As parameters are seen as meta-variables, different roles or viewpoints can be taken. HECK (2001) transfers the three variable roles placeholder, polyvalent name and variable object as described by FREUDENTHAL (1973) on to parameters, while URSINI & TRIGUEROS (2004) transfer their 3UV model onto parameters and distinguish between a parameter as a general number, unknown and variable.

DRIJVERS (2001), however, describes four different roles of a parameter: parameter as a placeholder, as a changing quantity, as a generalizer and as an unknown, while VAN DE GIESSEN (2002) only elaborates the first three. These four parameter roles according to DRIJVERS (2003) will be described in more detail, as they can give valuable hints, how students conceptualize parameters as well as where possible problems can occur.

Parameter as a Placeholder
The view of a parameter as a placeholder can result in consideration of specific values, one by one, for the parameter. A systematic change of the value is not intended and when a new value is chosen, it represents a new situation. Therefore it is a static view of the parameter. The aim is not to find the value of the parameter. Visualising the parameter as a placeholder in a graphic model, one graph would be replaced by another each time the parameter is changed. DRIJVERS (2003) describes this role as a ground level of understanding (p. 68 f). VAN DE GIESSEN (2002) states that this role is not suitable for solving problems.

Parameter as a Changing Quantity
The parameter as a changing quantity is connected with a numerical value, however a systematic change is intended. It is a dynamic view of the parameter. VAN DE GIESSEN (2002) describes it therefore as a kind of "sliding" parameter (p. 69), which runs through a set of values. Because of the second order structure of parameters, changing them influences not only locally but the global situation of all representations including the graphs. This is different compared to changing a variable, as that would act only locally. In a graphic visualization this role can be represented by a dynamically changing graph (DRIJVERS 2001).

Parameter as a Generalizer
A parameter as a generalizer is used to generalize over situations, e. g. for a family of functions. The parameter is not a single value, nor dynamically walking through a set, but rather the parameter is assigned a set of values at the same time (VAN DE GIESSEN 2002, p. 98). Similar to a family of functions DRIJVERS (2003) labels it as a "family" parameter. Visualising this parameter role in the graphic model, a family of functions can be represented by a sheaf of graphs (DRIJVERS 2003, VAN DE GIESSEN 2002). Both VAN DE GIESSEN and DRIJVERS state that a higher level of understanding is necessary for this role.

Parameter as an Unknown
The fourth role DRIJVERS (2003) identifies is parameter as an unknown, which in contrast to parameter as a generalizer is a particular case. This role effects the hierarchy between parameter and variable, as in most cases the variable is treated as the unknown, but while using parameter as an unknown it becomes the unknown-to-be-found (BILLS 2001, DRIJVERS 2003).

The different roles of parameters are on different levels of understanding (DRIJVERS 2003). The step from understanding parameter as a placeholder to understanding parameter as one of the other three roles needs an ontological shift also called reification by SFARD (1991, p. 19). Students then view expressions including parameters as a complete object (DRIJVERS 2003, p. 73). DRIJVERS (2003) develops a hierarchy between the different roles using the level concept according to HIELE. In this framework parameter as a placeholder is the ground level of concrete visual understanding. The parameter as changing quantity, generalizer and as an unknown, span a relational framework including the new objects parametric expressions and formulas, sheaves of graphs and classes of situations (DRIJVERS 2003, p. 71). Drijvers conjectures that parameter as a generalizer influences this the most, so it is positioned in the highest position (DRIJVERS 2003).

Relevance of the Concept of Variables and Parameters for the Study Presented Here
The concept of parameters will play a crucial role in the design of the material for
the intervention. I will follow the classfication of parameters according to DRIJVERS
(2003) as changing quantity, placeholder, generalizer and unknown and not as a role
of variable.

2.3.3   Misconceptions of Variables and Parameters

Students encounter a number of difficulties in learning the variable concept and
misconceptions can develop. These also can be applied on parameters, as parameters
can be seen as second-order variables or even as a variable role. Thus, it is not differed
between parameters and variables in this subsection, rather the misconceptions are
relevant for both concepts.

BILLS (2001) identifies the shift between the different meanings as a possible
obstacle while understanding variables and recommends drawing attention to these
shifts but the similarities and differences between variables, numbers and words
have also been described as a source of difficulties (WAGNER 1983).

A comprehensive overview of the misconceptions described in this section can
be found in Table 2.1.

Different Variables—Different Values
Some students believe that different variables must have different values, so the
variables in $3x + 4 = 5$ an $3y + 4 = 5$, must have different solutions (described
e.g. in WAGNER 1981, WARREN 1999, FUJII 2003). This has been described as non-
conserving strategies (WAGNER 1981). This misconception in a different direction
also occurs, if students do not realise that in one equation, where the variable occurs
more than once (so e.g. $3x + 4 = 5x + x$), it must have the same value (FUJII 2003).

Numerical Value on Alphabet Rank
As variables are nearly always represented by letters, the misconception that the
value of the variable is attributed to the position in the alphabet so $a = 1, b = 2$
etc. has been identified many times (e.g. by WAGNER 1981, WARREN 1999, SPECHT
2009). This might lead to situations in which the next consecutive integer after x
might be identified as y (WAGNER 1983).

Tasklabel
Students might think that with the expression $3a$ the part a of task three is meant
(WARREN 1999).

Table 2.1 Misconceptions Described in the Literature Regarding the Variable Concept

Misconception	Description
Different variables—Different values	Students think that two variables cannot have the same value (WAGNER 1981,WARREN 1999, FUJII 2003).
Numerical value on alphabet rank	The position in the alphabet determines the value, so $a = 1, b = 2, \ldots$ (WARREN 1999, SPECHT 2009).
Tasklabel	If the variable follows a number so e.g. $3a$ it is seen as part a of task three (WARREN 1999).
Closure	Students think that a result cannot have any operation in them so they omit them (WARREN 1999)
Equal sign	The equal sign is often perceived as a prompt to compute (WARREN 1999).
Letter equal value	The variable is used as an abbreviation of the numeral so t for ten (SPECHT 2009).
Roman numerals	Students think the variable is a roman numeral (SPECHT 2009).
$x = 1$	Students think that $x = 1$, probably due to the convention that 1 does not need to be written in $1 \cdot x$ (MACGREGOR & STACEY 1997, SPECHT 2009).
Reversal mistake	Reversing the relation, so for example $6S = P$ instead of $S = 6P$ to describe the relation that for every six students there is one professor (ROSNICK 1981, DRIJVERS 2003, SPECHT 2009, AKINWUNMI 2012).

Closure

Some students have the need to have a single answer without arithmetic operations signs in them and, thus, might omit the operations signs in order to achieve this, e.g. $a + b = ab$ (WARREN 1999).

Equal Sign

There can also be a confusion with the equal sign as in arithmetic it is sometimes used as a prompt to compute, whereas in algebra the two sides of the equation are equal (WARREN 1999, p. 314).

Letter Equal Value

Students in Grade 4 often show the misconception according to the rank in the alphabet as described above or assume that the variable is the first letter of the value, e.g. t for ten (SPECHT 2009).

Roman Numerals
Some students perceive the variable as roman numerals, so x would be always ten (SPECHT 2009).

$x = 1$
While students in Grade 8 often interpret x standing for the value of 1, SPECHT (2009) interprets this as a misunderstanding of the rule that the multiplication sign, for example in $1 \cdot x$ can be omitted (SPECHT 2009, p. 137 f.). This mistake is also identified by MACGREGOR & STACEY (1997), who conjecture that it might also be caused due to the power of x being 1 when no power is written.

Reversal Mistake
Many authors (e.g. SPECHT 2009, AKINWUNMI 2012, DRIJVERS 2003) have described the reversal mistake, which often occurs in the student-professor-problem described by ROSNICK (1981). In the student-professor-problem an algebraic term is asked for to describe the relation that for every six students there is one professor. The main mistake is to reverse the relation and state that $6S = P$, when S stands for number of students and P stands for number of professors, instead of the correct term $6P = S$ (ROSNICK 1981).

Relevance for This Thesis
Students operate with variables while working in the intervention, so the presence of misconceptions is possible. Especially the problem of $x = 1$ might pose an obstacle during the intervention, this might hinder insight into functions.

2.4   Representations and Change of Representations

Representations can be defined as "something that stands for something else" (PALMER 1978, p. 262; DUVAL 2006, p. 103). Due to this definition it implies the existence of two worlds: the *represented* and the *representing* world (PALMER 1978, KAPUT 1985). With this it is important to clarify what the representing and represented worlds are, and what aspects of the represented world are modelled through which aspects in the representing world and their correspondence (PALMER 1978, p. 262, see also KLINGER 2018).

Representations are an essential part of mathematics as no mathematical object can be accessed without the use of representations. This is a huge difference to other scientific disciplines where even though representations are used heavily the regarded objects are mostly still accessible through perception or instruments

(DUVAL 2006, p. 107). This situation leads to a paradoxical situation and problem for understanding which,

> "at each stage of the curriculum, arises from the cognitive conflict between these two opposite requirements: *how can they distinguish the represented object from the semiotic representation used if they cannot get access to the mathematical object apart from the semiotic representations?"* (DUVAL 2006, p. 107 emphasis in the original)

It can be differentiated between *external* and *internal representations*. While external representations can be physically seen, e.g. a sketch or a table, internal representation are only mentally and in mind. Both play a relevant role as communication can only be achieved with the help of external representation, but internal representations are needed to operate mathematically (e.g. KAPUT 1985, HIEBERT & CARPENTER 1992).

DUVAL (2006) identifies four representation registers as a part of semiotic representation which play a role in mathematical activities. Important factor is that one can switch from one register to another without losing any information about the representant. The four registers described by him are natural language, the symbolic-algebraic register in which one can operate following certain rules, cartesian graphing and visual like figures in elementary geometry like a drawing of a triangle (DUVAL 2006, p. 110). The change between representations is often seen as a "critical threshold for progress in learning" (p. 107) in order to counter the paradox of learning described above and as in each register different aspects of the mathematical object become more prominent.

He distinguishes two kinds of transformations regarding representations: treatments and conversions. Treatments occur completely within one register, for example solving an equation, whereas with conversions one transforms from one register of a mathematical object into another, for example graphing a function from its equation. Thus, conversion poses a more complex transformation than treatments as "any change of register first requires recognition of the same represented object between two representations whose contents have very often nothing in common" (DUVAL 2006, p. 112).

With both transformations of registers different problems in understanding can occur. Performing treatments within a multifunctional register, e.g. visualization of congruency in elementary geometry, are often complex and specific. For example in geometry students need to notice the relevant parts of a figure to be able to argue correctly for many kinds of task (DUVAL 2006, p. 116 f.).

The second problem for understanding presented by DUVAL (2006) concerns the conversion between register. Problems with conversion might limit the potential for learning. Representation registers on which the conversion is applied are the source representation and conversion results in a target representation. Source and target representation can be converted into each other in a congruent or non-congruent way. In a congruent conversion, the information represented can be mapped one-to-one, while in a non-congruent conversion this is not necessarily the case (DUVAL 2017, p. 88). For example converting the statement "the set of points whose ordinate is greater than their abscissa" (DUVAL 2006, p. 112) to the symbolic expression $y > x$ is a congruent conversion, while the conversion of the statement "the set of points whose abscissa and ordinate have the same sign" to the symbolic expression $x \cdot y > 0$ is a non-congruent conversion (DUVAL 2006, p. 113, DUVAL 2017, p. 88). A congruent conversion of $x \cdot y > 0$ would be "the product of abscissa and ordinate is greater than 0" (DUVAL 2006, p. 113). Students also might face difficulties when source and target representation are switched. Students find it considerably easier to graph a linear function from its equation than matching given graphs to equations (DUVAL 2006, p. 122). DUVAL (2006) states as "the root of trouble in mathematics learning: the ability to understand and to do by oneself any change of representation register" (DUVAL 2006, p. 122). To be able to understand what is relevant in a representation students need to investigate how variations in the source register and target register connect. For Duval a comprehension of mathematics only starts when the same mathematical object is recognized in at least two registers This is one argument in favour of using multiple representations simultaneously.

In the case of functions, four typical representations[21] have often been described in the literature: situative-verbal, graphical, symbolic and tabular (JANVIER 1978, KLINGER 2018). A situative-verbal representation of a function is usually a verbal description, JANVIER (1978) even includes diagrams, pictures and experiments in this category (JANVIER 1978, p. 3.4). Often the situative-verbal representation includes the real-world context. The graphical representation would be the plot of the function, while the function table would be a tabular representation. For an overview of activities to be undertaken while changing between representations see Table 2.2.

As described above, DUVAL (2006) postulates the need of at least two representations for understanding. Over the years, using multiple representations has been found to be important in the teaching and learning of all mathematical concepts, with KAPUT (1992) also describing the change between different representations as a part of "true mathematical activity" (p. 524). He even proposes the use of linked representations in which a change in one representation immediately transfers to

---

[21]I will use the term representations in the following in the sense of Duval's representation registers

Table 2.2 Examples of Activities During a Change of Representation. Adapted from JANVIER 1978, p. 3.2 and KLINGER 2018, p. 67

from \ to	situative-verbal	tabular	graphical	symbolic
situative-verbal	paraphrase, reduce	computing values	visualizing, sketch	modelling
tabular	reading, contextually interpreting a table	refining or enlarging a table, sorting	plotting a graph	interpolating, fitting
graphical	contextually interpreting a graph	reading values	transformation, zooming, graphically differentiate, integrate	curve fitting, interpolating
symbolic	interpreting formulas through parameter recogniton	computing	sketch	treatment, differentiate, integrate

the other as *"the cognitive linking of representations creates a whole that is more than the sum of its parts"* (KAPUT 1989, p. 179 emphasis in the original). KAPUT (1992) described that this linking can help students to visualize the connections between representations in a different way than a static change of representation and this leads to a deeper understanding. The linking of representations can be supported through the use of technology (FERRARA et al. 2006) and some authors even describe this dynamic linking of different representations as essential for developing a full understanding of concepts (e.g. THOMAS 2008; DUNCAN 2010). What purposes multiple representations can have and if they should be linked and represented simultaneously will be presented in more detail in section 3.2. However, linking the representations is not the only benefit of using technology. An overview of the benefits as well as risks will be given in chapter 3.

Use of Technology 3

Over the last decades a number of different tools, software etc. have been introduced into the mathematics classroom. These tools can now perform nearly every mathematical operation taught in secondary school and some can even be used for diagnostic purposes. New theoretical approaches for the analysis of learning processes have been developed or adapted from other theoretical approaches in order to explain learning processes when using technology. One theory explaining the processes during the use of technology is the *instrumental approach* (section 3.1). When using technology in teaching and learning, it has to be specified when it is appropriate and even beneficial using it (KAPUT 1992). In order to determine benefitting circumstances, chances and risks of using technology need to be taken into account. Using technology might even evoke certain misconceptions (section 3.2). During the last fifty years, there have been a number of studies searching for evidence for the benefit of using technology and the empirical results regarding their use are discussed in section 3.3. Using technology while learning about quadratic functions and specifically their transformations has also been a focus of a number of studies (section 3.4). Finally, using technology can also be seen as some kind of communication and might even have an influence on interactions between students in the classroom (section 3.5).

3.1   Artifacts and Instruments: The Instrumental Approach

A number of theories have been used to explain influences of technology use over the years (SINCLAIR & YERUSHALMY 2016). One often utilized approach—namely the *instrumental approach*—will be presented here since the study presented in this thesis investigates the influence of technology on a concrete learning process. The

L. Göbel, *Technology-Assisted Guided Discovery to Support Learning*, Essener Beiträge zur Mathematikdidaktik, https://doi.org/10.1007/978-3-658-32637-1_3

instrumental approach describes the processes taking place while working with e.g. digital tools (DRIJVERS & TROUCHE 2008).

The instrumental approach was adapted for the use of technology (see for example VERILLON & RABARDEL 1995, GUIN & TROUCHE 1999, ARTIGUE 2002, MONAGHAN et al. 2016) and distinguishes between an *artifact* and an *instrument*. DRIJVERS & TROUCHE (2008) define an artifact as

> "the 'bare tool', the material or abstract object, which is available to the user to sustain a certain kind of activity, but which may be a meaningless object to the user as long as that person does not know what kinds of tasks the 'thing' can support in which ways" (DRIJVERS & TROUCHE 2008, p. 367).

For example, a cello would be a meaningless object for a non-musician. In the context of the instrumental approach digital tools are artifacts which can be used for different purposes. Only when developing psychological constructs while using the artifacts, they can become instruments through the process of *instrumental genesis* (TROUCHE 2005a). Therefore, students need to work with an artifact in order to acquire skills and develop so called *schemes*. An instrument could be described by the equation

$$instrument = artifact + schema$$

(DRIJVERS & TROUCHE 2008, p. 368). Illustrating this using the Cello example, the cello only becomes an instrument (in the sense of the instrumental approach) when "a meaningful relationship between the artifact and the user for dealing with a certain type of task" (DRIJVERS & TROUCHE 2008, p. 367) exists and the user intends to solve the task. So for the cello, the meaningful relationship for playing e.g. a certain piece of music would develop through playing and practising. Therefore, developing mental schemes needed to play (how to hold the bow, how to grasp the cello, the coordination of both hands, playing single notes then whole phrases and finally the complete music piece).

A digital tool like a handheld calculator in the hand of a small child is also a meaningless object. Only when learning how to use it through practising and learning the syntax needed to operate it as well as learning the mathematical operations, mental schemes are developed and the handheld calculator can be used as an instrument.

When working with artifacts in order to develop instruments there are two processes taking place. Processes acting towards the artifacts are called *Instrumentalization* while processes acting towards the student (sometimes called subject in the descriptions of the instrumental approach) are called *Instrumentation*. These two kinds of processes are components of the bi-directional process of instrumental genesis.

Instrumentalization describes the process of the students using their preferences and conceptions to adapt the way the artifact is used and can lead to altering of it (DRIJVERS & TROUCHE 2008). It shows what the user believes to be the intended use of the tool and therefore can lead to "enrichment of the artifact, or to its impoverishment" (TROUCHE 2005a, p. 148).

In the cello example it could lead to tuning of the cello a halftone lower to play older pieces which were composed for a different concert pitch. The instrumentalization process can be divided into three different stages: *discovery, personalization* and *transformation* (TROUCHE 2005a, p. 148). In the discovery stage one gets to know the artifact and selects the relevant functions, while in the personalization stage one adapts the tool to one's hand and in the transformation stage one transforms the tool e.g. adapting a toolbar or programming (TROUCHE 2004, DRIJVERS & TROUCHE 2008). For digital tools like computer algebra systems, the discovery stage could consist, for example when trying to solve an equation, of students selecting the relevant function, e.g. the solve command. If the CAS allows it, students would then possibly put the solve command on a shortcut key to personalize it. In the transformation stage it might even be possible that they programm a different solve command so for example the solutions are given as fractions or decimals as wished.

The three stages can also be illustrated using the cello example. The discovery stage can be represented by discovering different tunings of the cello and how the position of the fingers result in different notes. In the early stages of learning to play a cello the cello is then often personalized by using glue dots to visualize the positioning of the fingers, removing them again once the player is proficient enough. In the transformation stage the cello could then be tuned to a different concert pitch in order to play certain pieces.

Instrumentation is directed towards the student. The possibilities and constraints of the artifact influence the way in which the schemes are developed (DRIJVERS & TROUCHE 2008, p. 368 f.). Returning to the cello example, the given pitch restricts the pieces a cellist could play, for example a piece written for a violin cannot necessarily be played by a cellist, but the four strings open up the possibility of playing chords. These possibilities and constraints influence the mental scheme of the cellist, a piece written for a violin would not be chosen to play for example.

The same applies to digital tools. A dynamic geometry system can only be used to some degree to solve equations. A computer algebra system would be better suited for this purpose.

TROUCHE (2004) (see also TROUCHE 2005a) identifies three different types of constraints: *internal, command,* and *organization constraints* (TROUCHE 2004, p. 290). Internal constraints determine what the user cannot change, for example the resolution of the display. Command constraints are linked to the syntax of commands which needed to be inserted, e.g. some tools require parentheses for sin(2), some do not, while organization constraints concern the available information and structure and are often dependent on designer choices (TROUCHE 2005a, p. 147).

Instrumentation and instrumentalization are not always exclusive and cannot be clearly identfied in learning processes. Therefore only instrumented action schemes as described below are visible (TROUCHE 2004, p. 295).

The schemes developed by the user during the instrumental genesis are often called *utilization schemes* and can be differentiated into *usage schemes* and *instrumented action schemes* (TROUCHE 2005a). TROUCHE considers instrumented action schemes as a set of usage schemes. Usage schemes can be simple actions (for example turning on handhelds) and are aimed at managing the artifact (TROUCHE 2004, p. 287). Instrumented action schemes are developed to complete specific tasks, for example computing a limit (TROUCHE 2004). Constraints and possibilities of the artifact affect the schemes that are built while using the artifact (TROUCHE 2004, p. 290). Therefore, similar artifacts with only slightly different constraints and possibilities can lead to different instruments as the mental schemes are different. The same applies to the same artifact being used by two different students with the result that this artifact can also develop into different instruments and instrumented techniques (TROUCHE 2005a, p. 158 f.). If the same student uses the same artifact at different times with different intentions it can also develop into different instruments. Multirepresentational systems (MRS) are always a collection of artifacts and can and will be developed into a number of instruments using different aspects of the MRS (TROUCHE 2014, p. 311).

But are students able to conduct their instrumental genesis without some kind of guidance how to conduct it? This seems unlikely and, therefore, the term *instrumental orchestration* describes the guidance of students' instrumental genesis as well as the coordination of the various and numerous processes of instrumental genesis going on in the classroom, which have to be guided and coordinated. Going back to the example of a cello, instrumental orchestration has often been symbolized by an orchestra conductor (DRIJVERS & TROUCHE 2008). This methaphor is not without problems, however, as it implies a hierarchical top down instrumental genesis through the teacher. In later articles it is often substituted or supplemented by the metaphor of a Jazz band leader, who recognizes and uses the instrumented techniques developed by the students, but also guides different musicians to produce a good sound (TROUCHE & DRIJVERS 2010). DRIJVERS & TROUCHE define instrumental orchestration as "the intentional and systematic organization of the various artifacts available in a computerized learning environment by the teacher for a given mathematical situation, in order to guide students' instrumental genesis"

(DRIJVERS & TROUCHE 2008, p. 377). Two components have been identified as relevant for instrumental orchestration, namely how the different artifacts are combined to an artifactual environment, so called *didactic configurations* and their *exploitation modes* (TROUCHE 2004, p. 297). Exploitation modes describe how a didactic configuration is used (DRIJVERS et al. 2010). DRIJVERS et al. (2010) add a third component the so called *didactical performance* involving the decisions and changes taken by the teacher while using the didactic configurations. This matches the jazz band leader who reacts to the improvisations of the musicians. The orchestration can act on three different levels.: "level of the artifact, (...) level of an instrument or a set of instruments, (...) level of the relationship a subject maintains with an instrument" (TROUCHE 2005b, p. 211).

One often used example of instrumental orchestration is the second-level instrumental orchestration through a *Sherpa-at-work*, in which a student operates a calculator that is projected overhead. The student can then be used as a guide hence the term Sherpa, which is a guide in the Himalaya (TROUCHE 2004). The projecting of this one student work can give insight into his/her instrumented action schemes. Exploitation modes in this didactic configuration can be, for example that only the sherpa's calculator is on or every calculator is used or only paper-pencil work (TROUCHE 2004).

3.2 Affordances and Constraints of Using Technology

"It is one thing (...) to use a tool, but quite another to use it effectively." (PENGLASE & ARNOLD 1996, p. 85)

PENGLASE & ARNOLD state the above citation in their review on the use of graphic calculators. But what makes a tool use effective? Over the last decades a vast number of chances and risks of using technology were identified. The ones most relevant for the scope of the study will be presented following a classification of RUCHNIEWICZ & GÖBEL (2019), for other affordances and constraints one can refer to the large number of thematic survey or reviews (e.g KAPUT 1992, PENGLASE & ARNOLD 1996, FERRARA et al. 2006, ZBIEK et al. 2007, BARZEL 2012, DRIJVERS et al. 2016). Most aspects presented here can be seen either as affordances or constraints, depending on the goal of the teaching as well as the integration into the learning.

Offloading Procedural Parts
The use of technology can reduce the procedural parts of mathematics and hoping to increase conceptual understanding. It has been shown in a number of studies

that this indeed can be the case (e.g. KIERAN & DRIJVERS 2006, BARZEL 2012). The question remains, though, which procedural parts should be delegated to technology and which should not (ARCAVI et al. 2017, p. 107 f.). Critical voices fear a substantial loss of pen and paper skills. TALL et al. (2008) compared twelve studies, which investigated the link between procedural skills and technology use and concluded that there is no difference in the development of procedural skills (TALL et al. 2008, p. 239). CAVANAGH & MITCHELMORE (2000a) amongst others identified a risk offloading procedural parts as students tended to accept the output of technology, even if their intuition told them the results were wrong.

Linking Multiple Representations
The use of multiple representations is crucial for learning about concepts of mathematics (see section 2.4). As described above it was even proposed to use linked multiple representations. These multiple representations can have *complementary*, *constraining* or *constructing* functions (AINSWORTH 1999, 2006, 2014). Depending on the function, a linking between the different representations seems essential.

- Complementary functions: Two or more representations can have a complementary function when the information represented differs or the representations support different processes (AINSWORTH 2006, p. 188). An example would be using tables, equations and graphs in a simulation (AINSWORTH 2014, p. 467). While tables can offer insight into specific points of the simulations the graphs can present a broader picture and enable students to draw conclusions about the general situation. Presenting relations between the different representations is not required, they can be interpreted on their own (AINSWORTH 2014, p. 467).
- Constraining functions: A representation can be used to constrain the interpretation of another representation and in due course support understanding. One example would be using an animation to support a dynamic graph (AINSWORTH 2014, p. 467; for an example see the self-assessment tool in RUCHNIEWICZ & BARZEL 2019). It is necessary to translate between the different representations to utilize the constraining functions. The change between the different representations can be linked dynamically.
- Constructing functions: The last function identified by AINSWORTH illustrates a support of understanding through the use of multiple representation, e.g. connecting a velocity-time and a distance-time graph to deepen the understanding of functions (AINSWORTH 2014, p. 467). For deeper understanding, the different representations need to be related to each other, but this is often a difficulty (see also section 2.4).

This linking of representations can be supported through the use of digital technologies (FERRARA et al. 2006). KAPUT (1992) coined the term of a "hot link" (KAPUT 1992, p. 530), for the situation where a change in one representation immediately leads to a change of another representation. There are, however, critical voices against using multiple representations simultaneously. The multiple representations might be overwhelming and overburdening for students due to higher cognitive load (BOERS & JONES 1994), others argue that students prefer one representation over another and therefore providing more than one can offer the students a choice (AINSWORTH 2006).

Quick Visualization
Technology can be used to visualize easily very quickly and statically as well as dynamically (e.g. HEUGL et al. 1996; BARZEL et al. 2005, p. 39). ZBIEK et al. (2007) identify the risk of confusion by the student, e.g. if a graph is replaced by another on the screen by typing in a new function equation. The timing of the new graph appearing can lead to contradicting information. Suppose a graph is displayed and a new function equation is entered into the calculator. Then the graph and equation are not the same function and thus contradicting each other. In the view of ZBIEK et al. (2007) this would be a breach of mathematical fidelity (ZBIEK et al. 2007, p. 1177).

Dynamic Visualization
Newer technological tools have the possibility of directly visualizing changes on mathematical objects, for example parameter variation through sliders or even direct dragging (VOLLRATH & ROTH 2012). This could foster the developing of the covariation Grundvorstellung (see for example DOORMAN et al. 2012). SEVER & YERUSHALMY (2007) conclude that the dynamicity guides the students to look at functions as an object and the direct manipulation of graphs helped some students, while others preferred using sliders. The use of sliders could avoid the contradictory display of information as described above, however it might then lead to students being more occupied with moving the slider and not realising that they are conducting a mathematical activity (ZBIEK et al. 2007, p. 1177). The dynamic change might also lead students to superficial examination of constructs without reflecting their observations (DRIJVERS 2004, BARZEL & GREEFRATH 2015).

Fostering Discovery Learning
The use of technology can foster discovery learning in a number of ways. The above described aspects (offloading calculation, fast visualizaiton, dynamicity) play a role in the fostering discovery learning through the possibilities to generate a

great number of examples while students are investigating a mathematical object. Students can search for structures and connections through the possibility of generating a number of examples (BARZEL & GREEFRATH 2015). But the possibility to explore many examples in a short time might also lead to unreflected actions and not necessarily conceptual understanding (BARZEL et al. 2005). As with any use of technology, a reflective discussion of the generated examples is needed and might then lead to the discovery of new mathematical concepts (DRIJVERS 2000). The use of technology during discovery learning emulates to some degree the process of experimental mathematicians (as described by SINCLAIR 2004, BORWEIN 2005, BORWEIN & BAILEY 2008). For discovery learning the four uses of technology "to gain insight and intuition, (…) to produce graphical displays that can suggest underlying mathematical patterns, (…) to discover new patterns and relationships, (…) to test and especially falsify conjectures" (SINCLAIR 2004, p. 235) are relevant.

Perception and Influence of Technological Settings
GOLDENBERG (1988) identifies scale as a focal issue for the use of technology. Technology enables students to change the scale easily, but they must understand the influence of scale on the output. MITCHELMORE & CAVANAGH (2000) also identify scale as an obstacle as well as the problem that students did not connect the zoom function with a change of scale. The use of technology for graphing might also emphasize the graph as an object and not so much as a set of individual points (GOLDENBERG 1988, p. 167). This would mean that the use of graphing software would emphasize the Grundvorstellungen object and not the Grundvorstellung mapping (see also subsection 2.1.2).

Apart from the scale obstacle, MITCHELMORE & CAVANAGH (2000) (see also CAVANAGH & MITCHELMORE 2000a, 2000b, 2000c) identified three main categories of difficulties while working with graphics calculators: *accuracy and approximation, linking representations*, and *representation by pixels* (MITCHELMORE & CAVANAGH 2000, p. 263 f.). The accuracy and approximation difficulties occur on a numerical level when students associate more decimal places with more accurate numbers but preferred integers. The linking representations difficulties stem from not connecting the graphs with the symbolic representations and accepting the given graph without reflecting on it. Therefore, students were seemingly not confused when the graph of a quadratic function seemed to be a straight line (MITCHELMORE & CAVANAGH 2000). With older graphic calculators the pixels which made up the graph were clearly identifiable and could lead to difficulties interpreting the graphs, because students did not realise the way the graphic calculators produced the graphs (MITCHELMORE & CAVANAGH 2000, p. 264). Since computers and handhelds now have the possibility of higher resolutions than in 2000, the pixel representation problem described by MITCHELMORE & CAVANAGH might not be relevant any longer.

These obstacles and illusions are important to notice and overcome, but they are not cause enough for abandoning using technology in class completely especially as there are a number of affordances of the technology described in the literature. But is there quantitative evidence that technology has a beneficial influence on students learning? The next section will outline some broader empirical results regarding this influence.

3.3 Influence of Technology Use on Students' Learning

The number of chances and risks are abundant and a large amount of studies have been conducted on the effect of technology use over the last 50 years. In order to achieve some kind of overview, there have been a number of systematic meta-studies and literature reviews regarding the influence of technology use on students' learning (e.g. PENGLASE & ARNOLD 1996, LI & MA 2010, BARZEL 2012, CHEUNG & SLAVIN 2013, for a non-systematic review see for example SCHMIDT-THIEME & WEIGAND 2015, DRIJVERS et al. 2016, LINDMEIER 2018). The effect can be viewed on two levels. To find out if technology has a statistically significant effect on the learning, one should look at the quantitative meta-studies. However, a more qualitative approach is needed to find out why technology might be effective (DRIJVERS 2018).

The quantitative results are overall not very convincing (see DRIJVERS et al. 2016) and some voices say that "there is little solid evidence that greater computer use among students leads to better scores in mathematics and reading" (OECD 2016. p. 145). Not all voices are so critical though. For example RONAU et al. (2014) state that technology can improve student understanding but do not harm computational skills. DRIJVERS (2018) reviewed current meta-studies and literature reviews extending one part of a topical ICME-Survey by DRIJVERS et al. (2016) and found more open questions regarding the *"if, how,* and *how much* to use digital tools" (DRIJVERS 2018, p. 162, emphasis in the original). Taking five meta-studies from the last decade into account (namely LI & MA 2010, RAKES et al. 2010, CHEUNG & SLAVIN 2013, STEENBERGEN-HU & COOPER 2013, SOKOLOWSKI et al. 2015), he found varying effect sizes, which were only small to moderate. Taking a closer look at the differences, he identfied five possible factors and resulting conclusions for the differences in the average effect sizes (DRIJVERS 2018):

– Student age and level: DRIJVERS (2018) states that younger students profit more than older ones (p. 166).

– Mathematical domain: Geometry or basic mathematics seem to be suited better
  for the use of technology than algebra (DRIJVERS 2018, p. 167).
– Learning goals and teaching style: Teaching styles with higher-order learning
  goals profits more than others (DRIJVERS 2018, p. 167).
– Duration and sample size: Interventions shorter than one year seem more effective, while effect sizes in small size studies are larger. But single-teacher interventions are not more effective than whole-school interventions (DRIJVERS 2018,
  p. 167).
– Development over time: There has been a rapid development of new technologies
  over the last decades. One can conjecture that this would lead to greater effect
  sizes, but the meta-studies show no such effect (DRIJVERS 2018, p. 167).

It can be concluded that the success of using technology relies on the implementation, the technology design and the orchestration by the teacher. LI & MA (2010)
showed that technology used in constructivist approaches has a stronger effect on
mathematics achievement than in other more teacher-centred approaches (LI & MA
2010, p. 228).

Gaining insight into which implementation and technology design benefits learning processes can be achieved through conducting qualitative studies. These are usually not included in quantitative meta-studies, but provide valuable information on
e.g. mediating effects of digital technology and generate new theoretical constructs
for the learning (DRIJVERS et al. 2016, HEID 2018). Overall, it can be concluded
that using technology "is a help, but not a breakthrough" (CHEUNG & SLAVIN 2013,
p. 102) and there are many factors like the implementation, tasks and teaching styles
influencing the results that have to be taken into account while planning the use of
technology.

But for this thesis it is important to take a closer look at how function transformation can be enhanced through the use of technology, as the study presented here
will investigate transformation of quadratic functions.

3.4   Learning Function Transformation With the Use of Technology

A number of studies have tried enhancing the learning of function transformations
through the use of technology.

One very early study using a multi-representation software was conducted
by BORBA (1993) (see also BORBA & CONFREY 1996). Using the framework of
prototypic functions, multiple representations and transformations developed by

CONFREY & SMITH (1991) and the program *function probe* as a multi-representational software, BORBA (1993) investigated a different approach to the learning of vertical and horizontal stretches and translations as well as reflections of absolute value, quadratic and step functions (BORBA 1993, p. 10 f.). With this software, students were able to manipulate a graph of a function not only by changing the table or equation entries, but also by dragging it with the computer mouse (BORBA 1993, p. 6). Students started with visualization of graphs then investigated the connection between graphs and tables, and only then focussed on the connections of graphs and algebraic representations (BORBA & CONFREY 1996, p. 320). BORBA also implemented the metaphor of *rubber sheets* first coined by GOLDENBERG & KLIMAN (1988), which can be used to explain influences of vertical and horizontal stretches to a graph and refines it to a *double rubber sheet*, where the graph is drawn on one rubber sheet and the cartesian plane on another (BORBA 1993, p. 31). A horizontal stretch could then be conducted by pushing or pulling in the left/right direction of the graph rubber sheet, while keeping the rubber sheet with the cartesian plane still (BORBA 1993, p. 32). This double rubber sheet metaphor was used by one of the students to explain the direction of a horizontal translation. The student, however, conducted the stretch on the back rubber sheet with the cartesian plane on it and then shifted both rubber sheets, so that the graph was back in its original place (BORBA & CONFREY 1996, p. 329). The series of teaching experiments with two students showed potential of his approach, however it was also suggested that the benefits of this approach depend on the students' personal preference and strength (BORBA 1993, p. 344).

Using an exploratory learning environment to foster conceptualization of functions transformations, students used a multi-representation program on computers in which algebraic expressions could be entered and they were then graphed (GADOWSKY 2001). The focus of the study was on how the technology was used, thus the used expressions were also recorded. The students were asked to alter the standard parabola and identify as many transformations as possible. As with other studies in the case of transformation, the students had more difficulties performing a horizontal transformation on the quadratic functions than performing a vertical transformation. All students attempted a vertical translation of $f(x) = x^2$ to $f(x) = x^2 + c$ and to $f(x) = ax^2$, while only five percent attempted a horizontal translation like $f(x) = (x - b)^2$ (GADOWSKY 2001, p. 71 f.). The participating students were also able to make some generalizations of their findings through testing hypotheses, which was supported by using the technology after overcoming some technical problems in the first part of the study. This exploring and testing was supported by the inquiry nature of the task as it appealed to the students' curiosity (GADOWSKY 2001).

Intending to foster the concept of linear transformations on functions SEVER & YERUSHALMY (2007) also used inquiry tasks supported by technology. Students used tools with which they could transform graphs by directly manipulating through dragging. They identified two benefits, namely "immediate and dynamic images" and the "direct and sensual manipulations of graphs" (SEVER & YERUSHALMY 2007, p. 1517) that supported the conceptualization.

Likewise using possibilities of immediate dynamic visualization, while focussing on the influence of parameters on quadratic functions amongst others, MCCLARAN (2013) used interactive GeoGebra applets and a guided exploration to facilitate understanding of vertical translation (so $f(x) = x^2 + c$) and vertical compression and expansion (so $f(x) = ax^2$). However, they did not drag directly at the graph as the students in the study by SEVER & YERUSHALMY (2007), but manipulated the parameters through sliders. The students were then prompted to view certain values for the parameters and complete function tables of it. Apart from the sliders for manipulating the parameter, the interactive applets consisted of a standard parabola, the transformed function equation and graph as well as a linked function table of both the standard and transformed parabola. They also included a slider for x, which controlled a vertical connection of the y-values both of the standard and transformed parabola for the x-value input through the slider. Participating classes were split into a treatment and control class. The control class was taught the influence of the two parameters through the teacher, while the treatment class investigated the influence through use of the applets. Comparing pretest to posttest scores, the treatment class outperformed the control class. This difference stems from an increase in the procedural understanding of the students in the treatment group (MCCLARAN 2013).

Like MCCLARAN (2013), PIHLAP (2017) compared an approach to teaching quadratic functions traditionally with teaching it supplemented with technology and designed three computer-assisted lessons. While the control group was taught the concept of quadratic functions over 15 lessons traditionally. In the treatment classes three lessons were planned for the treatment to be held in computer labs, working on exploration of quadratic functions guided through worksheets and supported by GeoGebra. In the first computer-based lesson the students explored the influence of a, b, c in the standard-form through sliders and guided questions, whereas in the second computer lesson students were prompted to plot specific examples, manipulate these through dragging and note connections between the graph and equations. In the last computer-based lesson students created patterns using quadratic functions. All classes completed a pre- and posttest. Even though no statistical differences in the learning outcomes of control and treatment classes could be observed, students in the treatment classes had a more positive attitude towards the topic and half of the participating students thought that using computers helped their understanding (PIHLAP 2017). The lack of statistical difference in the learning outcomes might be attributed to the small number of lessons in which the computer was used. The overall positive attitudes towards the technology supports the assumption that computers can facilitate learning of mathematical concepts (PIHLAP 2017).

The research into function transformation using technology shows promising results regarding the effectiveness of the technology use. Splitting the influences into seperate investigations regarding vertical and horizontal transformation seems necessary to induce a deep investigation into all of them. If this is not implemented, horizontal transformations are investigated less (see GADOWSKY 2001). Using inquiry tasks seem appropriate to motivate students for their investigation and produce more positive attitudes, but the learning outcomes show varying results (see GADOWSKY 2001, PIHLAP 2017). Regarding the kind of technology used, classic function plotters (as used by GADOWSKY 2001) as well as the special options of drag mode (used by SEVER & YERUSHALMY 2007) and sliders (used by MCCLARAN 2013 and PIHLAP 2017) show potential for the investigation, but it could be conjectured that they are not all equally beneficial. This will be investigated in the study presented here.

Students necessarily interact with the technology and other students while working with technology. Thus, learning without communication is nearly impossible and learning with technology can foster discussions and communication as well as influence the kinds of communication that occur. This connection will be explored deeper in the following section.

3.5   Communication and Technology

While learning there is a need to communicate, therefore DRIJVERS et al. (2016) and BALL & BARZEL (2018) took a closer look at communication while learning with or through technology. They state that all types of technology have an influence on and/or foster communication. Also, the affordances of general-purpose vs. specific-purpose technology differ and both are complemented by a number of online communication tools e.g. audience response systems and offline e.g. interactive whiteboards. They see one affordance of technology in the possibility to share displays and students' work. While DRIJVERS et al. (2016) differ between communication through and of technology, BALL & BARZEL (2018) distinguish between three types of communication while working with technology: "communication *through* technology, *with* technology or *of* technology displays" (BALL & BARZEL 2018, p. 233, emphasis in the original). They define the three aspects in the following way:

- "Communication *through* technology involves use of technology to support face-to-face communication or communication between students and/or teachers who are not in the same location.
- Communication *with* technology considers the entry of syntax, selection of menu items, programming or any command that drives the technology to produce a display. This communication is, for example, through key strokes, by touching a screen, using gestures to move objects on the screen, or by providing verbal commands.
- Communication *of* technology displays is evident when a technology display is a stimulus for discussion. This discussion could occur in a range of contexts, for example, through two students' consideration of one shared screen or through public display of student work via technology such as an interactive whiteboard or a data projector."

(BALL & BARZEL 2018, p. 233 emphasis in the original)

The three types can appear at the same time. BALL & BARZEL (2018) give the example that the last two aspects (communicating with and of) can be seen while entering syntax. The three aspects and their influence on learning will be presented in the following.

Communication through Technology
Communication through technology can be observed in class, in virtual communication between peers of one class e.g. via social networks and in cross-cultural communication across the world (BALL & BARZEL 2018). Communication in class can be evident when a screen is displayed and prompt discussion, for example through a "sherpa student" (as defined in GUIN & TROUCHE 1999 and described in section 3.1). Newer technology available nowadays even gives the possibility to collect screens of all students or distribute data for further use. Due to this there might be more time available for classroom discussion. BALL & BARZEL (2018) state the potential of technology use to change students' way of thinking about mathematics and foster collaboration and communication (BALL & BARZEL 2018, p. 235).

Communication with Technology
Even though some sociologists claim that it is no communication, BALL & BARZEL (2018) follow Schneider (2007 as cited in BALL & BARZEL 2018) and Peschek (2007 as cited in BALL & BARZEL 2018) in classifing entering syntax or using a touchscreen as communication with technology. While communicating one needs to know the specific language or commands and the roles for example variables can play in computer algebra systems (see for example HECK 2001). The resulting displays of the input need to be deciphered and therefore can foster reflection and metacognition (BALL & BARZEL 2018, p. 237).

Communication of technology
With communication of technology, also "communication of technology displays" (BALL & BARZEL 2018, p. 233), students or teachers use the display as a starting point to discuss with other students or teachers and, therefore, might further conceptual development. These starting points can be of different kind, for example an unexpected result and posing a cognitive conflict factor which fosters metacognition. BALL & BARZEL (2018) propose that through these kinds of communication the technology scaffolds learning.

It can be conjectured that in the course of working in pairs with technology a combination of *communication with* and *communication of* will occur, if students explain or show something to each other, while interacting with the technology. In the following this kind of communication with active use of the technology will be identified as *Communication including technology*.

Relevance for This Thesis
All four kinds of communication while using technology play a role in the study presented in this thesis, however to varying degrees. Communication through technology is only present to a very limited degree as students do not use the technology as support for the communication. Students work in pairs and are supposed to use the technology to communicate about the topic. Communication with technology occurs while students use the technology to display graphs of functions for investigating the connections. Using the technology as a starting point for discussion (e.g. about counter-intuitive position of graphs) would be a situation where communication of technology surfaces. Finally, the ideal case of using the technology to explain or discuss something can also occur during intervention and therefore communication including technology can also be present.

Discovery Learning

4

Guidance can be defined "as any form of assistance offered before and/or during the inquiry learning process that aims to simplify, provide a view on, elicit, supplant, or prescribe the scientific reasoning skills involved" (LAZONDER & HARMSEN 2016, p. 687). One kind of inquiry learning, where this guidance can be implemented is discovery learning where students are asked to find out the core of the topic themselves. Over the past 60 years the method of discovery learning has been discussed and researched on multiple times. One main advocator for using discovery learning was Jerome Bruner, while David Ausubel opposed discovery learning in an outspoken way and instead promoted more guidance (HERMANN 1969, see section 4.1). In the German-speaking community Heinrich Winter and his book on discovery learning (WINTER 1989) promoted discovery learning and is still relevant which is shown by reactions by mathematic educators in the third edition of his book (WINTER 2016, p. xi–xvi). The research findings are ambiguous as the studies show no clear picture about what kind of discovery learning is beneficial to learning. As a consequence of the research findings, different types and grades of guidance in the discovery process have been described (section 4.2). Taking the presented arguments into account a working perception of guided discovery will be developed (section 4.3).

Another kind of guided discovery is scientific discovery, which has been used to simulate the process of research in the classroom. The learners follow five set stages (orientation, hypothesis, experimentation, conclusion, evaluation) to discover new concepts (DE JONG & VAN JOOLINGEN 1998, GÖSSLING 2010). This has been shown to be also beneficial while learning mathematics (PHILIPP 2012). However, for this thesis we will examine a different kind of guided discovery, which will be more open and not as predetermined as the five stages which have to be followed in scientific discovery which lead to a cycle of experimentation.

© The Author(s), under exclusive license to Springer Fachmedien Wiesbaden GmbH, part of Springer Nature 2021
L. Göbel, *Technology-Assisted Guided Discovery to Support Learning*,
Essener Beiträge zur Mathematikdidaktik,
https://doi.org/10.1007/978-3-658-32637-1_4

4.1   Introduction to Discovery Learning

A number of different definitions of discovery learning have been formulated over the years. Discovery can be defined as the "matter of rearranging or transforming evidence in such a way that one is enabled to go beyond the evidence so reassembled to additional new insights" (BRUNER 1961, p. 22). However, it can also be formulated that "discovery learning occurs whenever the learner is not provided with the target information or conceptual understanding and must find it independently and with only the provided materials" (ALFIERI et al. 2011, p. 2). The two definitions present the concept of discovery with two different foci. ALFIERI et al.'s definition describes more what the teacher has to consider, if students are asked to discover. The target of the procedure is to discover the unknown goal and for this materials are provided. BRUNER's definition targets the concept from the process side, as he focusses on what students have to do with the given evidence.

The degree of how much is to be discovered varies in the definitions, however, the common element in the definitions is that the "target information" (ALFIERI et al. 2011, p. 2) is to be discovered.

There has been an intensive discussion of the arguments for and against discovery learning in the last sixty years. Some arguments for and against discovery learning will be presented in the following, taking only the original texts of BRUNER (1961) and AUSUBEL (1963, 1964) into account.

Even though BRUNER (1961) describes an example in which discovery learning without any guidance, e.g. through materials, hints or strategies, is not productive (BRUNER 1961, p. 31 f.), he also hypothesizes four main benefits of discovery learning namely increasing intellectual potency, shifting from extrinsic to intrinsic motivation, learning the heuristics of discovery and the aid to memory processing. These will be explained in more detail in the following.

- *Increasing intellectual potency:* While learning by discovery, learners need to organize the discovered information, which will make it more accessible later when problem-solving (BRUNER 1961, HERMANN 1969).
- *Shifting from extrinsic to intrinsic motivation:* As well as learning to organize information while discovering, the students also become their own "paymaster" (BRUNER 1961, p. 28) and learn to motivate themselves. Therefore, the need for extrinsic motivation decreases as intrinsic motivation increases. AUSUBEL (1964) opposes and states that reception learning can also motivate the students. Reception learning is "the situation where the content of the learning task (what is to be learned) is presented to rather than independently discovered" (AUSUBEL

1963, p. 1). AUSUBEL assents, however, to the fact that discovery learning may lead to intrinsic motivation.

- *Learning the heuristics of discovery learning:* BRUNER (1961) hypothesizes that only through exercise in learning by discovery the learners pick up on the heuristic of discovery, which then can be used in other tasks. To this argument AUSUBEL (1964) disagrees with this hypothesis and states that these heuristics are not transferable to other disciplines and therefore not helpful to learn:

> "Also, from a purely theoretical standpoint alone, it hardly seems plausible that a strategy of inquiry, which must necessarily be broad enough to be applicable to a wide range of disciplines and problems, can ever have, at the same time, sufficient particular relevance to be helpful in the solution of the specific problem at hand" (AUSUBEL 1963, p. 158).

- *Aiding intellectual potency:* The last benefit BRUNER (1961) states, concerns the learners' memory. He conjectures that discovering on their own leads to more accessible storage of the material in the brain. AUSUBEL (1963) discredits this benefit with rejecting the presented experiment of children learning word lists as a situation in which no discovery needs to take place.

AUSUBEL argued that there was no empirical proof for the benefits and proposed that students should first be able to "respond meaningfully, actively, and critically to good expository teaching" (AUSUBEL 1964, p. 302) before educating them as problem-solvers. He does not explain what he means by expository teaching, but a definition can be found in BRUNER (1961), who defines it as "the decisions concerning the mode and pace and style of exposition are principally determined by the teacher as expositor; the student is the listener" (BRUNER 1961, p. 23).

Reception learning and discovery learning can be described as two ends of a continuum with the amount of guidance as a differentiating factor (AUSUBEL et al. 1978). While reception learning is providing complete guidance, no guidance at all would require complete discovery learning. They name the instructions with a grade of guidance in between those two contrasting positions *guided discovery* and attribute greater effectiveness to guided discovery than to pure reception or pure discovery learning (AUSUBEL et al. 1978, p. 335 f.).

All approaches to learning in general and discovery learning in particular need to take the current capability of the learners into account. Here VYGOTSKY's (1978) *Zone of Proximal Development* (p. 85 f.) gives an appropriate frame to face and overcome this challenge.

The Zone of Proximal Development (ZPD) is the

> "distance between the actual development level as determined by independent problem solving and the level of potential development as determined through problem solving under adult guidance or in collaboration with more capable peers." (VYGOTSKY 1978, p. 86)

The Zone of Proximal Development is a widely used concept to describe development of children in a number of areas e.g. reading, writing, science and mathematics (CHAIKLIN 2003). It enables a more individualised view on learning, which can in turn give valuable insight into how discovery learning can succeed as the discovery learning needs to take the prior knowledge of students into account.

Some authors refer to the Zone of Proximal Development when learning with technology (e.g. MONAGHAN et al. 2016). It can even be argued that the adult or peer interaction required by VYGOTSKY (1978) can be taken over by technological tools or other artefacts by realising communication with technology (BROWN & CAMPIONE 1994). Thus students working with peers while using technological tools might benefit through the guidance of both the peer and the technological tool.

The discussion if discovery learning is beneficial or not led to the conducting of a number of (meta-)studies which analysed different versions of discovery learning. The studies presented here include one meta-study and two literature reviews (namely MAYER 2004, KIRSCHNER et al. 2006, ALFIERI et al. 2011) and are often cited as they describe the difficulties regarding the efficacy of discovery learning.

MAYER (2004) reviews the literature comparing pure discovery with guided discovery while teaching three different topics: problem-solving rules, conservation strategies and programming concepts (MAYER 2004). With pure discovery he characterises discovery methods with little or no guidance from the teacher, while with guided discovery he denominates methods, where the teacher "provides hints, direction, coaching, feedback, and/or modeling to keep the student on track" (MAYER 2004, p. 15). Conservation strategies are strategies in the context of a concept that some object remains invariant despite changing its appearance. One example would be two glasses of different forms and water poured from one glass into the other. Young children do not realise that the water still has the same volume, while children who have developed conservation strategies realise the invariance. While he agrees that a constructivist approach to learning is sensible, he argues that high behavioral activity methods do not necessarily lead to active learning. He states that "students need enough freedom to become cognitively active in the process of sense making, and students need enough guidance so that their cognitive activity results in the construction of useful knowledge" (MAYER 2004, p. 16). Therefore, it makes sense that guided discovery methods outperform pure discovery methods, as guided discovery meets both criteria (freedom for cognitively active and enough guidance) for active learning. Accordingly, he proposes abandoning pure discovery (MAYER 2004).

Taking an even broader view, some authors claim that all minimally guided instruction is less efficient and less effective than strongly guided instructional approaches for novice and intermediate learners (KIRSCHNER et al. 2006, p. 83). They state that discovery learning, problem-based learning, inquiry learning, experiential learning and constructivist learning are more or less equivalent approaches, which infer that students need to experience the methods of the particular discipline in order to learn. They then claim this assumption is erroneous. KIRSCHNER ET AL. also express that the working memory, which is severely limited in the amount of information it can process, will be unable to process all the new information encountered in minimal guidance instruction. Thus, the cognitive load connected with minimally guidance is so great that learning is highly unlikely. Even for advanced learners with prior knowledge, who therefore have strategies to process the new information, strongly guided approaches seem to be more appropriate (KIRSCHNER et al. 2006).

These limitations of discovery learning have been tested and discussed broadly, a systematic meta-analysis was conducted by ALFIERI et al. (2011). Their definition of discovery learning states the absence of a given target allowing an individual way. They chose 164 studies and conducted two main comparisons, which can serve as an operationalization of unassisted and enhanced discovery. Firstly they considered studies in which types of unassisted discovery learning were compared with more explicit instruction (see Table 4.1), for this they chose studies which compared some kind of unassisted discovery learning (first column in Table 4.1) with some kind of comparison conditions (second column in Table 4.1). So for example a study could compare unassisted discovery learning with worked examples. Their results show that unassisted discovery learning led to less learning achievement than more explicit instruction (ALFIERI et al. 2011. p. 7).

In a second comparison they considered studies in which enhanced discovery methods in particular generation, elicited explanation, guided discovery are compared with other types of instruction (see also Table 4.2). This second comparison showed that enhanced discovery learning (so learning that included some enhancement e.g. feedback or scaffolding) led to greater learning than the comparison conditions (ALFIERI et al. 2011, p. 7). But the analysis also showed that the generation method, where students react to questions by an experimenter, is less beneficial than the other two.

Adults benefited in a greater way than children from enhanced discovery, while adolescents benefit the least in all comparisons. They conjecture that this might

Table 4.1 Different Methods of Instruction Considered in First Comparison by ALFIERI et al. (2011)

Unassisted Discovery learning in first comparison	Comparison conditions
Unassisted: no guidance, learning through trial and error or practice problems	*Direct teaching:* topics are taught directly e.g. through lecturing
Invention: students invent own strategies or design own experiments	*Feedback:* any instructional design with direct feedback
Matched probes: minimal hints and feedback through probe questions, which are also provided in the comparison group	*Worked examples:* students are presented with worked examples
Simulation: computer-generated simulations to manipulate by the students	*Baseline:* students were not given the instruction of the discovery group, but completed other tasks in the same time
Work with a naive peer: students worked in equal learning pairs	*Explanations provided:* Learners were given the necessary explanations
	Others: other instructional condition not classified by other instructions

Table 4.2 Different Methods of Instruction Considered in Second Comparison by ALFIERI et al. (2011)

Enhanced Discovery learning in second comparison	Comparison conditions
Generation: learners "generate rules, strategies, images or answers to experimenters' questions" (ALFIERI et al. 2011, p. 5)	*Direct teaching:* topics are thought directly e.g. through lecturing
Elicited explanation: learners explain the information or material	*Worked examples:* students are presented with worked examples
Guided discovery: "instructional guidance (…) or regular feedback" (ALFIERI et al. 2011, p. 5)	*Baseline:* students were not given the instruction of the discovery group, but completed other tasks in the same time
	Explanations provided: Learners were given the necessary explanations
	Others: other instructional condition not classified by other instructions

be due to tasks which cater more to adult learners and therefore are not in the adolescents Zone of Proximal Development (ALFIERI et al. 2011). The findings indicate therefore that the guidance needs to take the zone of proximal development of the students into account if it is to be beneficial (ALFIERI et al. 2011, p. 12).

They conclude from their two comparison in their meta study that enhanced discovery tasks, which require "learners to be actively engaged and constructive seem optimal" (ALFIERI et al. 2011, p. 13). This can be achieved by implementing guided tasks with scaffolding, tasks requiring the learners to explain which is then followed by timely feedback or including worked examples. These kinds of instruction are described by MAYER (2009) as "active instructional methods" which lead to "active learning" (MAYER 2009, p. 192). How these kinds of assistance should be implemented needs to be empirically investigated (ALFIERI et al. 2011, MAYER 2009).

4.2   Guided Discovery

The empirical results of many studies imply that enhancing discovery learning through some kind of scaffolding, worked examples or feedback is necessary to achieve successful learning (ALFIERI et al. 2011, p. 13). But what are the criteria of different ways to guide students in their discovery? It can be differed into types and grades of guidance.

Different types of guidance have been described in the literature. It can be distinguished between *directive* and *nondirective support* (DE JONG & NJOO 1992). As the term directive support implies, it helps the learner directly to reach the goal and steers them towards it. For example as when students are learning a procedure, directive support would be intervening as soon as the students start going wrong. Another example of directive support can be direct hints at what the learner should do next or better such as "it is better to change only one variable at a time" (DE JONG & NJOO 1992, p. 422). Nondirective support however, provides no such direction, but can be provided for instance through a hypothesis scratchpad in which the students can input and save any hypothesis they might have. This nondirective support facilitates organisation of the learning process and therefore reduce the working memory load. Accordingly, it helps achieve what students would have done without a ready learning environment (DE JONG & NJOO 1992, DE Jong 2005).

In the case of scientific discovery, three different types of support have been described: *interpretative support, experimental support* and *reflective support* (REID et al. 2003). These kinds of guidance differ in what kind of activity it is supposed to support instead of the directiveness of support. Interpretative support helps the students to rearrange their knowledge or in generating hypotheses, for example the learners are presented with feedback regarding their interpretation. Experimental support helps the students conduct scientific experiments and interpret the emerging results, e. g. scaffolding through hints to apply the control of variables strategy (as defined in CHEN & KLAHR 1999) or guiding questions. In contrast, reflective support scaffolds the students' meta-cognitive skills (REID et al. 2003, DE Jong 2005).

The types of guidance described by DE JONG & NJOO (1992) and REID et al. (2003) are classified in relation to different frameworks, however DE JONG & LAZONDER (2014) and LAZONDER & HARMSEN (2016) describe different types depending on the directness of the guidance. They differentiate between *process constraints, status overviews, prompts, heuristics, scaffolds* and *explanations*. All six could be classified as directive support (as described by DE JONG & NJOO 1992), but vary in the directiveness.

– *Process constraints* are used to diminish the choices the learner needs to make during the discovery process. They are intended to "be used when students are able to perform the basic inquiry process but still lack the experience to apply it under more demanding circumstances" (DE JONG & LAZONDER 2014, p. 375). Thus, they are the least explicit type of guidance and can be implemented for example through splitting one complex task into more comprehensible subtasks. An example would be in the investigation of linear functions to first only present students with functions of the form $y = m \cdot x$ and introducing the y-intercept after a certain amount of time, compared to directly presenting the students with the form $y = m \cdot x + b$.
– *status overviews* or as DE JONG & LAZONDER (2014) call them *performance dashbord* are used to give students insight into their learning progress. The students are free to use it or ignore the information given through this.
– *Prompts* "are reminders to carry out a certain action or learning process" (DE JONG & LAZONDER 2014, p. 376). They are more explicit than the first two types of guidance as they are timed reminders to do a specific action, but give no specific instruction how to do this (LAZONDER & HARMSEN 2016).
– *Heuristics* provide this insight into how to perform the recommended action, hence they are more guided. There are different ways to present such heuristics, for instance by providing a full list of heuristics at the start of the inquiry process or by giving them at the appropriate times as timed cues (LAZONDER & HARMSEN 2016). An example of task-independent heuristic aids are so-called strategy keys, where problem-solving heuristics are presented to the students on keys during their problem-solving process (HEROLD-BLASIUS 2019).

– *Scaffolds* can be used to "provide students with the components of the process and thus structure the process" (DE JONG & LAZONDER 2014, p. 377). The use of scaffolds conforms with the Zone of Proximal Development as they take over the challenging parts and therefore assist the learner in reaching the learning goal. Scaffolds are intended to be removed, once the learner reachers the skill intended, but this is often not implemented (LAZONDER & HARMSEN 2016).

– *Explanations* are the most guided form described by LAZONDER & HARMSEN (2016). They are intended to be used for "learners who are (largely) incognizant of the action and how it should be performed" (LAZONDER & HARMSEN 2016, p. 689). Explanations can be given before or during the inquiry process (LAZONDER & HARMSEN 2016).

Deciding what type of guidance will be used directly influences the grade of the guidance during the inquiry process in general and therefore also in the process of discovery learning. The above presented classifications are not mutually exclusive, for example can experimental support as described by REID et al. (2003) be used as scaffolding but also as prompts when only guiding questions are given.

Apart from types of guidance it can also be differed into different grades, so to what degree the students are guided. In the case of discovery learning, BAROODY et al. (2015) differentiate between *highly guided, moderately guided, minimally guided* and *unguided discovery*:

– They understand *highly guided discovery* as "well-structured and moderately explicit instruction and practice" (BAROODY et al. 2015, p. 94), which includes scaffolding e.g. through strategically organized tasks and feedback so the learners are led to the target of the discovery without stating it.

– In *moderately guided discovery* the learners receive some degree of scaffolding and feedback, so for example "sequentially arranged items to underscore a relation" (BAROODY et al. 2015, p. 94).

– The approach of *minimally guided discovery* contains the same elements of moderately guided discovery with a lesser degree of scaffolding.

– With *unguided discovery* no such scaffolding or feedback is given.

Some definitions of guided discovery even define it as a teacher-centred approach in which the teacher asks closely defined questions and adapts to the students' answers to reach the goal of the lesson (see for example KERSH 1962, MOSSTON 1972, HIRSCH 1977), the method is sometimes described as "socratic method", sometimes as direct teaching. This kind of teaching might look like discovery learning, however the narrowing of the questions is intended and no real discovery takes place (WINTER

1988). In his book about teaching styles MOSSTON (1972) attributes a full chapter on the concept of guided discovery in this sense. He defines it as a teaching style in which

> "the teacher guides the student through small, sequential discoveries until he discovers the *focus*, the goal, that the teacher has selected: The student is presented with a sequence of questions (or clues) so meticulously arranged that the student *always* discovers the correct response until the target (the focus) is reached" (MOSSTON 1972, p. 121 f, emphasis in the original)

MOSSTON (1972) then presents four design principles which the teacher needs to decide on to enable students achieving the discovery. These design principles are the following four and can be used as good guidelines for conducting discovery successfully, not only for the guided discovery described by him, but also when designing guided discovery environments in which discovery by the students does take place.

Focus of the Guided Discovery

The guided discovery needs to have a specific focus, so the process becomes a "convergent process that leads the learner to discover a predetermined target" (MOSSTON & ASHWORTH 2002, p. 214). This focus provides a goal and direction and may be one of many things, e. g. discovery of facts, of their relationships, of concepts or particular behaviour (MOSSTON 1972, p. 128).

Initial Phase or Starting Point

Once the focus is set, there is the need to decide how to start the guided discovery. MOSSTON (1972) states that it is necessary to create an environment that serves as initial motivation. This can be achieved for example through a general statement, a particular physical setting or involving the learners in an activity (MOSSTON 1972, p. 129 f.).

Sequence of Tasks

The teacher needs to be aware of the various sequences of tasks which could lead to the predetermined target. Another important element of guided discovery in both the question-led approach as well as a more open approach is "a story line that inherently engages the participant" (GERVER & SGROI 2003, p. 6). Both the sequence of tasks as well as the story line should provide a clear structure to facilitate the students' discoveries.

Size of Discoveries
To achieve discovery the size of needed discoveries needs to be adjusted to the subject and focus as well as the learner's sophistication (MOSSTON 1972). This concurs with the theory of the Zone of Proximal Development (VYGOTSKY 1978) presented above, because if the size of the discoveries needed to reach the goal of the discovery learning is too great the goal might move out of the students' Zone of Proximal Development and thus will not be reachable by the learner.

GERVER & SGROI (2003) identify one more design principle.

Chance for a 'Eureka'-Moment
Guided discovery lessons also need an "Aha! Component" (GERVER & SGROI 2003, p. 7), so at some point in the lessons the students realise that they discovered mathematical ideas through their own exploration.

4.3   Guided Discovery in This Thesis

Taking all above presented arguments into consideration, for this thesis Guided Discovery will be the following.

By *Guided Discovery*[1] a series of moderately guided tasks or questions is meant, including some degree of scaffolding (see BAROODY et al. 2015) which lead the learners on their own to a predetermined learning goal. The tasks provide some structure, so the topic is comprehensible for the students. However, no imminent correction or reinforcement is given by the teacher, the learners are given the opportunity to discover and evaluate their own conclusions. Students discover in pairs or groups of three as to provide peer-to-peer interaction, which can benefit the learning. To support the students directive and experimental support (see subsection 4.2) can be provided through the wording of tasks or additional hints. To design such a guided discovery the focus of it, the initial phase, the sequence of the tasks and the size of needed discoveries need to be taken into account. How this is implemented in the frame of transformation of quadratic functions will be described in chapter 7.

Following this perception of Guided Discovery might influence the possible benefits of discovery learning stated by BRUNER (1961). Even though the intrinsic motivation might still be evoked if the design principle of a Eureka Moment is followed, the learning the heuristics of discovery learning might be less apparent if

---

[1]In the following, if guided discovery in this here described sense is meant, it will be set in title case. If the broader term guided discovery as described in the literature is meant, it will be written in lower case.

guidance in the process is given. However, following the results of ALFIERI et al. (2011) the guidance should be implemented to achieve higher learning success. The guidance is intended to move the goal of the discovery process closer to or into the learners' Zone of Proximal Development.

Part III
Empirical Study

Research Questions

The learning of transformation of quadratic functions has be shown to be difficult to students, partly because of the problems regarding conceptualization of parameters (see subsection 2.2.2 and section 2.3). Technology, however, has promising potentials to support learning of mathematical concepts, especially through dynamic and multiple representations (see Chapter 3). Likewise, Guided Discovery appears to be an appropriate way to foster constructivist learning of mathematical concepts (see chapter 4). But what happens if one combines all three aspects, might technology-assisted Guided Discovery be a beneficial way to learn about the conceptualization of parameters of quadratic functions? Conceptualization of parameters of quadratic functions can serve as a blueprint for the parameter concepts in higher order functions, thus developing a profitable way of teaching is important and valuable. In order to evaluate potentials and risks of technology-assisted Guided Discovery, it is aimed to investigate the influence through a study with the following research questions. The main focus will be on the role of the technology and especially if different functionalities of technology influence the learning differently. Thus, the Guided Discovery approach will be the same and only the technology assistance differed. The main research question of the study presented here is:

How does technology-assisted Guided Discovery influence the conceptualization of parameters of quadratic functions?

The research question of the study presented here can be splitted into subquestions regarding the three aspects technology, content, communication. The subquestions are:

L. Göbel, *Technology-Assisted Guided Discovery to Support Learning*,
Essener Beiträge zur Mathematikdidaktik,
https://doi.org/10.1007/978-3-658-32637-1_5

Technology-related subquestions

– What influence of the different technology use can be identified on the summary sheets?
– What affordances and constraints of the technology are visible?

Content related subquestions

– What conceptualization of parameters takes place?
– How does the technology use influence the students' learning?

Communication-related subquestion

– Which kind of communication while using technology occurs during the intervention?

Overview of the Study

In order to answer the research questions presented in Chapter 5 the study was designed and conducted as a control group intervention design in order to give insight into different design elements.[1] The main emphasis of the study was the role of technology in the Guided Discovery, thus for the three experimantal groups three different features of technological tools were chosen. All three features are possible to implement into a number of different programs: standard function plotter, the possibility to drag directly at a graph and sliders. For the control group it was chosen to not use technological visualization at all.

The intervention (see section 6.3) was planned to be three times forty-five minutes. This was deemed to be an appropriate length of time to achieve some degree of conceptualization as well as being a length so that it could be integrated into usual classroom teaching. The groups were assigned on a class basis, so all students in one class would use the same technological visualization. This non-random assignment of all participating students might influence the results, therefore a pretest (see section 6.2 and section 7.1) regarding the abilities of all students was administered in order to collect baseline data. The complete classes were assigned to the four groups nearly randomly except in a case of two classes from the same teacher or two or more classes from the same school participated in which case the classes were assigned to different groups. Deeper insight into the complete design of the study including symbols for the four groups is represented in Figure 6.1. A description of the collected data will be given in Chapter 9.

---

[1] The study and selected results have also been described in GÖBEL 2017, 2018, 2019, GÖBEL & BARZEL 2016, 2019, GÖBEL et al. 2017, RUCHNIEWICZ & GÖBEL 2019.

© The Author(s), under exclusive license to Springer Fachmedien Wiesbaden GmbH, part of Springer Nature 2021
L. Göbel, *Technology-Assisted Guided Discovery to Support Learning*,
Essener Beiträge zur Mathematikdidaktik,
https://doi.org/10.1007/978-3-658-32637-1_6

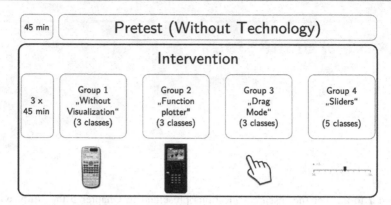

Figure 6.1 Overview of the Study

6.1   Decision of Technology Use

The three different features that were chosen can be implemented into a number of programs, e.g. Geogebra, TI-Nspire CX CAS, Cinderella. These three namely function plotter, drag mode and sliders were selected, as they represent different potentials of technology use (see section 3.2). The literature shows potentials for all three while discovering transformation of quadratic functions (see subsection 3.4 and e.g. GADOWSKY 2001, MCCLARAN 2013, PIHLAP 2017). But it can be conjectured that the three are not equally suitable for the purposes and therefore it was decided to compare them. As a program TI-Nspire CX CAS was chosen for a number of reasons. Schools in North-Rhine Westfalia usually choose either TI-Nspire CX (with or without CAS) or Casio CG-20 (or Casio Classpad as CAS-choice) as the graphic calculator used in the last two years of school, so teachers participating in the study had some experience with one of them. One of the groups was supposed to manipulate the graph by dragging, but the Casio Handhelds do not offer this feature. Additionally, the TI-Nspire is available as an App for iPads as well as handheld calculators. The iPad app offers the same features as the handhelds, however in some aspects it is more intuitive due to the multi-touch functions. TI-Nspire is a document-based system, which opens the possibility of using pre-programmed files to work on. Geogebra would have been also a valid choice, however the dragging feature to manipulate parameter a in $f(x) = a \cdot x^2$ is not possible in Geogebra. Also it would have meant organizing access to the computer labs at school, which presents another hurdle for running the intervention. In addition, students were asked to work in pairs to foster discussion, this is difficult working on handhelds so the decision was taken that TI-Nspire preferably on iPads was to be used. TI-Nspire was also used in teaching at the researchers university, so with schools who had not introduced a graphics calculator in Year 9, the researcher was able to supply the students with the technology for the intervention. It was decided that all students in the technological groups should use TI-Nspire CX CAS in order to ensure the same possibilities regarding hard- and software for all groups.

6.2 Pretest

To collect baseline data a pretest about preliminary knowledge was administered to all participating students. The baseline data would then be used to ensure comparability between the different groups. It was given to the students before the intervention, but after introducing the standard parabola in class. It was a 45 minutes test without help of technology. The decision to administer a test without calculator was based on the fact that the normally used calculators in class differed. Some classes had already introduced graphic calculators in Year 9 while others still used scientific calculators. This would have led to different conditions and therefore impaired comparability. The items covered a range of tasks, e.g. change of representations, solving equations. The topics were chosen with regard to the relevant topics in the intervention. The test was the same for all groups. The items of the test, their intention and skills needed for solving will be discussed in section 7.1.

6.3 Intervention

The intervention of the study was designed for the relatively short period of three times forty-five minutes, which is three standard-length lessons in Germany. This decision was made to offer teachers the possibilty of participating in the study, as it could be easily integrated into their normal teaching. Transformation of quadratic functions is taught in Year 9 in most states of Germany, which in North-Rhine Westphalia is currently the last year of lower middle school at the Gymnasium. Therefore teachers expressed their concerns that a longer intervention would cost too much time. At the same time it was also a logistic decision as only one class at the same time could participate due to access to the technology. But the short intervention might not be a big obstacle, as research has shown that even short interventions regarding technology use might be beneficial (see LI & MA 2010 and section 3.3).

The intervention was split into two parts, the first part taking 45 minutes, so one lesson, the second part conducted as a double lesson of 90 minutes[2]. In the first 45 minutes the students were given an introduction into the TI-Nspire system on Handhelds or iPads except for students in the control group who were given an introduction into function tables on their scientific calculators. This first introductory lesson was implemented to ensure that all students had worked with the media that was used in the second part of the intervention and therefore reduce the cognitive load while working with a new tool. It was hoped that, due to the intuitive handling of technology by students nowadays, by the second part of the intervention some grade of instrumental genesis (as described in section 3.1) had taken place and the students were able to use the tool as an instrument. In the introduction in the experimental groups, students were asked to work with TI Nspire CX CAS on a number of tasks concerning linear functions (see section B.1). They changed the window settings, displayed function tables and had one task specifically for the group the class was assigned to. In the *function plotter*[3] group the students plotted three linear functions trough inputting the function equation, while in the *drag mode* group the students transformed the function $f(x) = x$ through dragging. Finally in the *sliders* group students manipulated $f(x) = a \cdot x + b$ through two sliders for a and b. The introduction for the control group depended on the model of scientific calculator and was designed by the class teacher.

The second part of the intervention consisted of two lessons of 45 minutes which were conducted without a break. In this second part the students worked in pairs in a Guided Discovery as defined in section 4.3 on a series of tasks. The teachers role was only to enable a working atmosphere and if major problems occur to give some minimal hints to resolve them. The tasks were only presented through the worksheet described in subsection 7.2.3. The decision to use discovery learning with nearly no input through the teachers was made for a number of reasons. Using a method which does not rely on the teachers leads to more comparable results as teachers do have an immense influence. Also, teachers would then not be assessed in their teaching quality, which is an argument for participating. All teachers were given a timeline for the lessons, which they were instructed to follow as close as possible. If the teacher did not feel comfortable to conduct the lesson, the accompanying researcher in the classroom took over. It was decided that all students participating in the study should work on the same tasks to ensure maximal comparability. This might not

---

[2]Some schools had a lesson length of 67.5 minutes, in these cases the second part was shortened to this length.

[3]In the following, if the experimental group is meant, the technological feature is written in italics, if the feature itself is meant, it is set in typeset.

lead to maximum learning achievement compared to designing the task for each group specifically, however, the goal of comparing the different groups seemed to be worth the possible loss of learning achievement. The tasks for the second part of the intervention were designed in a number of steps and the developing of these will be described in section 7.2.

6.4   Different Intervention Groups

As described above, it was decided to implement three experimental and one control group. All four groups only differed in the kind of visualization the students had at their disposal. The reasoning behind the decision for the four different versions will be explained below.

Group 1: Without Visualization
Group 1 served as a control group compared to the three visualization groups. The control group in the intervention were allowed to use a scientific calculator without the possibilty of graphing functions. As the graphing calculator is only mandatory at Year 10 and upwards, this is the calculator mostly used by students in the Year 9 in North Rhine Westphalia. Therefore sketching graphs per hand is the way most students learn about transformation of quadratic functions. The version used in most classes of the intervention was the Casio fx-991 DE Plus, one class did not use a scientific calculator at all. This calculator can display function tables and the students were specifically allowed to use this advantage. However, the calculator cannot display any graphs, so the students would need to plot the values from the table themselves. Some critical voices of teachers regarding using technology favour this approach as students need to sketch the graphs per hand and therefore experience the differences of the graphs on their own, while sketching. Also this approach is a slow investigation, as sketching per hand is slower than only inputting the formula into a graphics calculator. It was conjectured that this might be a crucial point for the learning about the parameter concept.

Group 2: Function Plotter
Group 2 was the first of the three experimental groups and uses a static visualization. The group 2 students were given TI-Nspire Apps on Ipads and used it as a function plotter tool. They were able to plot as many functions they would like at the same time and display the function tables. Due to the programming the students were technically able to drag the functions, but if seen that they were dragging the function, the teacher or researcher discouraged this. Function plotter tools are widely available

and have been around for a number of decades. One benefit is that students may plot more than one function at a time and therefore directly compare two graphs. This group has a faster way of visualizing different graphs of functions than the control group through plotting them, however students in this group still need to actively decide what function to plot and then input the function equation, so it is still a slow investigation. To change the window of the graphed function the students could zoom in or out with two fingers, this gesture was already very intuitive for students as it is implemented on nearly every smartphone.

Group 3: Drag Mode
The second of the three experimental groups was the *drag mode* group, which uses a more dynamic approach. Students in the *drag mode* group were given TI-Nspire apps on iPads or TI-Nspire CX CAS handhelds and a pre-prepared file. This file included a graph of a standard parabola and a linked function-table and function equation. The students could drag the graph directly on the graph to a new position when dragging at the vertex point and also manipulate the value of parameter a by dragging at any other point of the parabola. Especially on the iPad App this feature is very intuitive to students. Therefore it was conjectured that this might be a beneficial kind of visualization. However, achieving integer values for the three parameters is more difficult than other visualizations as the drag mode allows all values. The possibility to observe the dynamically changing graph, equation and table simultaneous offers potentials that have already been described in the literature (see section 3.2). Students could immediately see the change in all representations and due to the dynamic change, there was at no time contradicting results on the display (which was one of the problems described in section 3.2).

Group 4: Sliders
The last of the three experimental groups was the *sliders* group, which also uses a dynamic approach. Like the students in the *drag mode* group, the students used TI-Nspire apps on iPads or TI-Nspire CX CAS handhelds and were given a pre-prepared file to work with. The file contained a graph of the parabola $a \cdot (x-b)^2 + c$, sliders to manipulate the value of a, b, c and a dynamically linked function table. The displayed function equation stayed the general form $f(x) = a \cdot (x-b)^2 + c$. The sliders could be manipulate in 0.1 steps and students could move the sliders from -10 to 10. These pre-set steps can serve as a process constraint (see section 4.2 and DE JONG & LAZONDER 2014) as the students only have a finite number of possible examples that can be displayed. The graph dynamically changed to the input value, so if a student moved one of the sliders, it was kind of a film of the changing graph. The conjecture was that, because students could manipulate each parameter on their

own, students would be able to conduct a specific investigation into the influence, while the benefit of dynamically changing with no contradicting results remains. Also both the *drag mode* and the *sliders* group implemented linked representations, which have been shown to be beneficial (see section 3.2).

6.5   Research Instruments

A number of research instruments were chosen and collected. All pretests for a baseline (see section 6.2) were collected, while in the intervention the summary sheets that were designed (see section 6.3 and section 7.2) were collected from all students and in all classes.

If the researcher was present during the intervention, two students per class were videographed in the second part of the intervention in order to gain insight into the processes that occur during the Guided Discovery.

Design of the Materials

In order to achieve insight into the research questions (Chapter 5) using the study described in chapter 6 different tasks had to be designed. For a baseline regarding preliminary knowledge in the field of functions and algebra, tasks for the pretest were designed (see section 7.1). Different skills were needed for solving the test tasks, which are described at each item. The Guided Discovery tasks for the second part of the intervention were developed in three rounds of design (see section 7.2). The different versions of the tasks for the intervention were tested and then re-designed for the goals of the study. Both the pretest items as well as the tasks for the interventions were designed and implemented in German, in this thesis English translations (by the author) will be presented in the text. The German versions of the items can be found in Appendix A and the German versions of the intervention tasks in Appendix B.

7.1   Design of Pretest Items

As described in section 6.2 a pretest of 45 minutes was run to achieve an overview of existing knowledge. Due to the legislation in Germany, teachers can decide in which order Year 9 topics are taught, therefore the test mainly contains items on Year 8 topics. The items covered a range of topics, namely i) functions and their Grundvorstellungen, ii) different representations and the change between them, and iii) the concept of variables. These topics are relevant for a full understanding of the parameter concept, which means that insufficient preliminary knowledge on these can directly influence the achievement in the intervention. The test also included a task on the parabola of the form $f(x) = a \cdot x^2$, its equation, properties and function table.

© The Author(s), under exclusive license to Springer Fachmedien Wiesbaden GmbH,   81
part of Springer Nature 2021
L. Göbel, *Technology-Assisted Guided Discovery to Support Learning*,
Essener Beiträge zur Mathematikdidaktik,
https://doi.org/10.1007/978-3-658-32637-1_7

For the purpose of establishing a knowledge basis as well as being able to compare the classes the test was corrected and then scored. A maximum score of 56 points could be achieved in the pretest. The scoring depended on the demands of the task, so for example noting coordinates of a given point would contribute one point to the final score, while solving the equation $6 \cdot a + 3 = 45$ correctly would contribute three points.

In the following, each of the items as well as the its purpose will be explained separately. The items for the pretest were partly taken from KLINGER 2018, other large-scale assesment tests (namely TIMS-study or VERA 8, a standardized test in Year 8 for all German schools) or were self-developed.

7.1.1   Finding Coordinates (Item P1ZC)

The self-developed item P1ZC Figure 7.1 focussed on the change from the graphical to the numerical representation. Students were asked to identify coordinates of four points on different graphs and identify x-values for a given y-value on the three given graphs. This item emphasizes the Grundvorstellung mapping (as described in subsection 2.1.2). The three functions chosen for this graph were two linear functions and a standard parabola. These three functions should be known by the students. The standard parabola was implemented especially for the second part of the item as it is the first non-constant function the students encounter that has two pre-images for a given function value.

A possible misconception for the first part of the item is the swapping of the x- and y-coordinates of the points. This is based mostly on students not knowing about the convention that the first coordinate is the coordinate for the independent variable. For the second part of the item a possible mistake would be not realising that $f2$ has two pre-images for the value of 1, while another mistake is computing $f(1)$ for the functions instead solving $f(x) = 1$.

For the final scoring of the tests, in part (a) each pair of coordinates was awarded one point, while in (b) for each function one point was awarded if pre-images of 1 were determined, so a total of 7 points could be achieved.

On the following graph you can see four points on the functions.

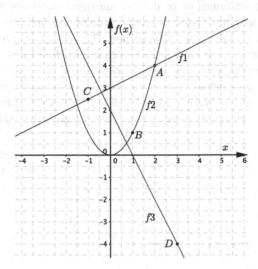

(a) Note the coordinates of A, B, C and D and write them in the given lines.

A (,   )
B (,   )
C (,   )
D (,   )

(b) At which position(s) x have the different functions a value of 1?

$f1$:
$f2$:
$f3$:

Figure 7.1 Pretest Item Finding Coordinates (P1ZC)

7.1.2   Algebra (Item U2PT)

The item U2PT was used for an overview of the students' skills regarding algebraic operation with variables and the German version was translated from the released items of the TIMS 2011 study (FOY et al. 2013, p. 57).[1] Students would need either to factorize the expression $2 \cdot a + 2 \cdot b$ to $2 \cdot (a + b)$ and substitute 35 into the equation or perform algebraic transformations on the first equation to substitute it into the second equation. Therefore, it focusses on treatments on the algebraic representation and the calculation aspect described in subsection 2.3.1. It can, however, also be solved by substituting values for a and b so the first equation is fulfilled and then simply compute the second equation, which would focus on the substitution aspect while seeing the variable as an unknown (see subsection 2.3.1). In both cases students would need a flexible way of variable uses. Possible mistakes would be miscomputing the value. For a correct answer three points were added to the final score.

Task: U2PT

$a + b = 35$. What is the value of $2 \cdot a + 2 \cdot b + 4$?

Answer: [   ]

Figure 7.2 Pretest Item Algebra (U2PT), FOY et al. 2013, p. 57

7.1.3   Variable as an Unknown (Item E4FV)

Item E4FV was Figure 7.3 similar to an example task for the standardised tests for year 8 in Germany (original version see *Fünfundvierzig* 2011). It focusses on the solving of equations, which is a major part in the algebra curricula. In the first part of the item the variable a would be needed to be treated as an unknown and treatments on the algebraic representations could be conducted to achieve a correct solution. This would then emphasize the calculation aspect as described by MALLE (1993). In the second part of the task students would need to show their way of solving. It was expected that many students would simply show the algebraic transformations needed to solve it, but it can also be solved by simply trying different values. A possible mistake would be simply dividing 45 by 6 without subtracting three first. Each of (a) and (b) contributed with a maximum of three points towards the score.

---

[1] The Figure 7.2 shows the original text from FOY et al. (2013)

(a) Determine the missing number a.

$6 \cdot a + 3 = 45$

Answer: [   ]

(b) Explain in detail what you did.

[   ]

Figure 7.3 Pretest Item Variable as an Unknown (E4FV)

7.1.4   Fill Graphs (Item N1FQ)

The item N1FQ Figure 7.4 was developed by KLINGER (2018, p. 243) and adopted for the use with the same purpose. Solving it correctly depends heavily on a working Grundvorstellung covariation as there are no exact values given, so pointwise checking of the graphs is not possible. Students would also need to convert between a verbal representation to a graphical one. KLINGER (2018) suggests that students might answer the item correctly using a graph-as-picture mistake (as described in CLEMENT 1985), as the correct answer is closest to the form of the swimming pool. A correct answer was worth three points.

Water is pumped consistently into the swimming
pool shown on the right.
Which of the graphs can represent this?

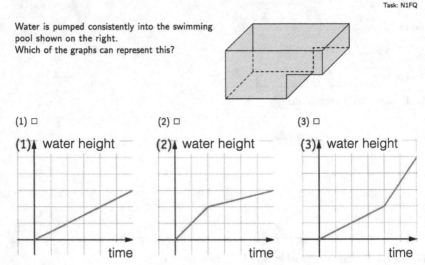

Figure 7.4 Pretest Item Fill Graphs (N1FQ). Translated from KLINGER (2018, p. 243)

7.1.5   Dependance of x and y (Item D3JG)

Solving item D3JG Figure 7.5 requires students to investigate the change of a function when the independent variable is altered. It was translated and reformulated
from DRIJVERS (2003, p. 363).[2] The item tests the understanding of covariation in
linear functions. It can be solved using the Grundvorstellung covariation, when the
students realise that the slope was negative and y decreases as x increases. It can
also be solved using the Grundvorstellung mapping first through substituting values
and then comparing the resulting values. As a distractor the function was not given
in the most common form $y = m \cdot x + b$ where m is the slope and b is the y-intercept,
but the y-intercept was put first in the right hand side of the equation. Therefore,
students would not immediately see that the slope is negative. For a correct answer
three points were given to the final score.

---

[2]The original in DRIJVERS (2003) states "Between x and y the relation $y = 5 - 2 \cdot x$ holds.
What happens to y when x gets larger?" (p. 363).

x and y are arbitrary numbers. It is $y = 5 - 2 \cdot x$. What happens to y, when we increase the value of x?

Figure 7.5 Pretest Item Covariation of Linear Functions (D3JG)

7.1.6   Variable Concept (Item D4JG)

The item D4JG Figure 7.6 was intended to evaluate the variable concept students showed. It was adopted from DRIJVERS (2003, p. 363) with the small change of switching a to x in $y = 15 + p \cdot x$ to resemble the form of linear function students are used to more closely. It was thought that this item might lead to identification of variable and parameter roles in the student answers, as students were asked to state a meaning for the variables x and y and the parameter p. It attributed in total 6 points to the final score.

Vermerk: D4JG

What do you think of when you look at the formula $y = 15 + p \cdot x$?

What could be the meaning of the letters y, p and x?

Figure 7.6 Pretest Item Variable Concept (D4JG)

7.1.7   Change of Representation (Item A6XY)

This self-developed item Figure 7.7 deals directly with the change of representation between graphical, numerical and symbolical representation of three functions,

Below you can see the function graphs, function equations and value tables of three functions.

(a) Assign each graph to a table and a term. Connect them with a line.

(b) Also, fill in the empty spaces in the function tables.

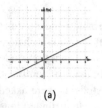   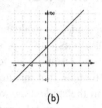   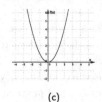

(a)   (b)   (c)

(1)

x	f(x)
-6	
-5	25
-4	16
-3	9
-2	4
-1	1
0	0
1	1
2	4
3	9
4	16
5	25
6	

(2)

x	f(x)
-6	
-5	-3
-4	-2
-3	-1
-2	0
-1	1
0	2
1	3
2	4
3	5
4	6
5	7
6	

(3)

x	f(x)
-6	
-5	-2.5
-4	-2
-3	-1.5
-2	-1
-1	-0.5
0	0
1	0.5
2	1
3	1.5
4	2
5	2.5
6	

(i)   (ii)   (iii)

$f(x) = \frac{1}{2} \cdot x$ $f(x) = x^2$ $f(x) = x + 2$

Figure 7.7 Pretest Item Change of Representation (A6XY)

two linear and one standard parabola. Change between representations of the same functions is a crucial step in understanding the function concept (see section 2.4). In part (a) connecting lines were to be drawn between the corresponding graph, table and equation. Students would need to change between the graphical and numerical representation and between the numerical and symbolic representation. This could be done in both directions. The two linear functions could be directly identified in the tables through their y-intercept. For each correctly drawn line, one point was added to the final score. The students were also asked to complete the missing values in the table. This was intended to show if they could calculate in their head or by using properties of functions without having a calculator. Calculating the values new for each x-value uses the Grundvorstellung mapping, while using the change in the values for the already displayed x-values would use more the Grundvorstellung covariation. Two values could be inferred through symmetry. The values that were to be calculated could not be read off the graphical representation as the graphs only showed the section for $x \in [-4, 5]$. For each correctly calculated value one point was added to the final score.

7.1.8   Family of Functions (Item Y3GJ)

This item Figure 7.8 was self-developed in order to determine, if students could identify the equation of a family of functions and therefore perform a change of representation between the graphical and symbolical representation. Solving this item would also require a certain understanding of the parameters in linear functions. The students were also asked to explain their decision to determine possible misconceptions. The distractors $f(x) = a \cdot x + 2$ and $f(x) = a \cdot x$ were chosen to represent one of the shown graphs, whereas the last distractor $f(x) = 2 \cdot x + 3$ represents a function equation with only x as a variable. To determine the correct equation students would need to realise that only the family of graphs represented by $f(x) = x + b$ has the same slope for all representatives of that family of functions, while the y-intercept can differ. Choosing the correct equation resulted in two points put towards the final score and up to three points for the explanation of their choice.

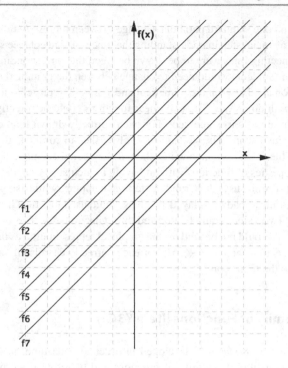

Which function equation can be used to describe all graphs?

☐ $f(x) = a \cdot x + 2$ ☐ $f(x) = x + b$ ☐ $f(x) = a \cdot x$ ☐ $f(x) = 2 \cdot x + 3$

Explain, why only your choice can describe all graphs.

Figure 7.8 Pretest Item Family of Functions (Y3GJ)

7.1.9   Properties of Linear Functions (Item R5TG)

Item R5TG Figure 7.9 was developed with the intention to test what properties of a general linear functions students would be able to identify. The general formula for a linear function in Germany is often given as $f(x) = m \cdot x + b$. In order to try to identify the actual understanding of parameters and not reproducing stereotypes, e.g. recollection of the role of m, the parameter was changed to a. All of these properties should have been covered in Year 8. Students would need to interpret the function equation and use their understanding of the linear function concept. The item also focusses on the change between the symbolic and verbal representations of properties. For each of the six statements, the students were asked to decide if the statement was true or false. As a typical misconception in the fourth statement it was tested if students had knowledge about reflections of linear functions. Reflecting a linear function $f(x) = a \cdot x + b$ at the x-axis would result in negating the complete equation and result in $f(x) = -a \cdot x - b$. All other statements are correct and these properties of linear functions should have been taught in Year 8, though some classes might be unfamiliar with the wording used. For each correct answer one point was awarded, wrong answers were not penalized.

Task: R5TG

Given the function $f(x) = a \cdot x + b$, where a and b are real numbers. Decide for every statement whether it is true or false.

	true	false
For $a = 0$ the graph is parallel to the x-axis.	☐	☐
b states the y-intercept.	☐	☐
For $a = 1$ und $b = 0$ it is the first angle bisector.	☐	☐
The function $g(x) = -a \cdot x + b$ looks like f, only reflected at the x-axis.	☐	☐
x in the function must be substituted for all numbers.	☐	☐
a and b represent numbers that are fixed in a given function.	☐	☐

Figure 7.9 Pretest Item Properties of Linear Functions (R5TG)

7.1.10 Properties of Quadratic Functions (Item R4TG)

This item Figure 7.10 was taken from KLINGER (2018, p. 301) and used to determine pre-existing knowledge of the students towards parabolas. For each of the five statements the students were asked to decide if the statement was true or false. Students need to connect the technical term standard parabola with the function equation $f(x) = x^2$. At the point of the pretest no transformation of parabolas through parameter a had been taught to the students, so this item was included in order to determine if students could either transfer their knowledge about linear functions to quadratic functions or had some other kind of pre-existing knowledge. It was expected that students would be able to answer the first statement correctly by realising that multiplying the variable x with 0 would result in a constant function and not a parabola. For the second statement students would need to realise that a can also be negative and x^2 is always positive and therefore the parabola could also open downwards. With the same reasoning students could infer the correct answer to statement three. If $0 < a < 1$ it is definitely positive and therefore the parabola open upwards. A problem in answering statement four could be that students do not comprehend that $f(x) = x^2$ is the same as $f(x) = 1 \cdot x^2$. Finally, students would need to have working knowledge of reflecting graphs at the x-axis and the resulting change in the function equation. Contrary to KLINGER (2018), the five statements were corrected individually and in total five points could be achieved, wrong answers were not penalized.

Task: R4TG

Given the function $f(x) = a \cdot x^2$, where a is a real number. Decide for every statement whether it is true or false.

	true	false
For $a = 0$ it is a standard parabola.	□	□
The parabola f is always opened upwards.	□	□
For $0 < a < 1$ is the parabola f opened downwards.	□	□
For $a = 1$ it is the standard parabola.	□	□
The function $g(x) = -a \cdot x^2$ looks like f, only reflected at the x-axis.	□	□

Figure 7.10 Pretest Item Properties of Quadratic Functions. Translated from KLINGER 2018, p. 301

7.2   Development of the Tasks for the Intervention

The intervention tasks are well-known and often used tasks in which students investigate the influence of parameters in quadratic functions (see subsection 2.2.2). It was adapted for the purposes of the study presented here in two cycles and empirically evaluated: a minimally guided approach, an approach implementing more scaffolding and the final version of enhanced guidance to realise the Guided Discovery approach. The design of the different versions and the evaluation of the tasks will be presented in the following[3].

7.2.1   First version: a Minimally Guided Discovery Approach

The decision was made that the vertex form $f(x) = a \cdot (x - b)^2 + c$ of quadratic functions was to be used. This has the advantage that the parameter b only has influence on the horizontal position of the parabola compared to the standard or normal form $f(x) = a \cdot x^2 + b \cdot x + c$, where b influences both the horizontal as well as the vertical position of the parabola. This was in accordance to the conventions in German schools as mostly the vertex form is taught in the context of transformations first. For the first version a minimally guided approach was adopted. This was chosen to test if the minimal guidance was already enough to invoke meaningful investigations of the influence without restraining the discovery process too much. Therefore, in the task students were given the vertex form and asked what influence parameters a, b and c have on the graph, this was chosen as the *focus of the discovery*[4] (as described in section 4.2 and by MOSSTON 1972). The students were supposed to compare the resulting graphs with the standard parabola, whose equation was also given (see Figure 7.11).

> **Task**
>
> Let a, b, c be arbitrary, real numbers. Investigate, how changing the paramters a, b and c in the function $f(x) = a \cdot (x - b)^2 + c$, $x \in \mathbb{R}$ will change the graph. Compare the graph to the graph of a standard parabola $g(x) = x^2$.

Figure 7.11 First Version of the Task for the Intervention

To reduce the *size of the necessary discoveries* and taking the *sequences of tasks* into account some additional scaffolding was also provided. However, it was on a separate sheet of paper including the allowed technological tools (see Figure 7.12). The scaffolding was intended to point the students in the right direction by inducing a stepwise approach and evoking hypothesizing and the testing of the conclusions taken by the students. This might lead to a *Eureka-Moment* as demanded by GERVER

---

[3]In this chapter translated versions of the tasks will be presented, original German versions can be found in Appendix B.

[4]All design principles described in section 4.2 will be set in italics if referenced to in this chapter.

& SGROI (2003). Possible difficulties were anticipated in the wording of the task, as some students might not be familiar with the technical terms in the task.

Instructions:

- Do it in steps.

- Look at different values for a, b and c.

- Then try to generalize.

- Can you prove your assumptions?

Group 1 (without visualization) Permitted technical tools: None. Sketch graphs for specific cases.

Group 2 (function plotter) Permitted technical tools: Function plotter "Grapher"
You can plot more than one function at the same time if you click on the plus-sign in the grapher and then new equation.

Group 3 (drag mode) Permitted technical tools: TI-Nspire CX software. Use the given file. You can drag the function with your mouse and also manipulate the opening.

Group 4 (sliders) Permitted technical tools: Geogebra 5. Use the given file. You can manipulate the function with the slider bars.

Figure 7.12 Provided Scaffolding and Allowed Technical Tools for the First Version

Testing of the Minimally Guided Discovery Approach
The task was tested with 8 students of Year 9 in pairs at a private upper secondary school. The students had learned about the topic four months earlier, so it was assumed that they would have at least some knowledge of the technical terms used in the task. It was decided to test the task with students who had already covered this topic, because it would lower the threshold of understanding. If students were not able to investigate and come to meaningful conclusions, it would be clear that students with no prior knowledge would face even more difficulties understanding. If the students were able to gain considerable insight, the processes would show the crucial points of the task. For this piloting each pair tested the version for one of the four groups planned. So two students worked on the task without technological visualization, two students with a function plotter, two with the drag mode and two with the sliders. However, contrary to the description in section 6.1 it was decided to use other software (namely the Mac OS Grapher and GeoGebra) than TI-Nspire for the *function plotter* and *sliders* group, so the students would not need an introduction

on how to work with the software. For the *drag mode* group it was decided to use the TI-Nspire Software as other software did not offer the complete potential of dragging directly at the graphs. All software was used on Laptops, in order to capture the screens for the analysis. The students worked for about 35 minutes on their own, while the researcher only intervened to help with possible technological difficulties and tried answering questions without giving too many hints. All four pairs were videographed and their work collected.

Evaluation of the Piloting

Even though the students already covered the topic, the task was too ambitious for the students, so it could be argued that the intended *size of discoveries* was too great. Observations by the researcher and reviewing the videos led to identifying a number of obstacles that arose in each group. These will be presented in the following and the resulting changes of the task discussed. The observed problems can be classified in three categories: language-related, technology-related and topic-related difficulties.

Language-related Difficulties

The students in all groups did not understand what an arbitrary, real number means as well as did not recognise the symbol $\mathbb{R}$ for the real numbers. They also did not know what a parameter is. As all students were from the same class, these terms might not have been taught, even though students should be able to work with real numbers at the end of Year 8 (MSW NRW 2007).

Implications for a change of the task: To avoid these kinds of difficulties it was decided to simplify the language as far as possible and to elaborate more on what the tasks were.

Technology-related Difficulties

Groups 2, 3 and 4 were supposed to use software while working on the task. This was not apparent to them at first and they needed to be encouraged by the researcher to use the software. This probably stemmed from the unusual way to work in math class, as students at this school did not have graphic calculators in Year 9. Also they did not realise that they were allowed to use the provided scaffolding.

Students in the *drag mode* and *sliders* group were given a programmed file, however, these students wanted to type in the functions one by one and had to be reminded to use the way intended in their group. This also can be explained with uncommon features of the programs. Students in the *function plotter* group used the software "grapher" on Mac, in the planning it seemed more intuitive than plotting functions with TI-Nspire CX CAS, but for the two students using this software it proved to be difficult. Again this could be due to unfamiliarity of the software and showed the importance of an introduction into any program that was to be used in the intervention.

Implications for a change of the task: The *focus of the guided discovery* that included the technology use was not clear enough. Students should therefore receive an introduction into the technology used, so they know what they are allowed and encouraged to use. Also the problems with the handling of the software should be addressed in an introduction or with help sheets in the intervention. The software "grapher" is not as suitable as thought before the testing of the open approach. This strengthened the decision to use TI-Nspire CX CAS with all experimental groups.

Topic-related Difficulties

The students in all groups were unsure what exactly they had to do in the task as the *starting point* or the *focus of the discovery* was not made explicit enough. After some time all students tried to investigate. The aim to investigate the influence proved to pose a number of difficulties. The group using no technology was not able to sketch a single graph correctly. Because the function equation was given as $f(x) = a \cdot (x - b)^2 + c$ the students thought they would need to choose a value for $f(x)$ as well as substituting values for x in $a \cdot (x - b)^2 + c$. This led to wrong graphs. Students working with the *function plotter* successfully plotted functions for different values of a, b and c, but they changed all parameters each time and this complicated making the connections. To correctly interpret the transformation of the parabola it is wise to compare each changed graph with the standard parabola. Even though the comparison was explicitly stated in the task it was neglected by the students.

One part of the instructions dealt with trying to find proofs or reasons for the findings stated, but these were too hard to find or the students did not try to find reasons at all. Students were asked to discuss and verbalize what they were thinking, but they often wrote something down without a verbal connection to what they were doing with the technology. This led to difficulties analysing their work.

Implications for a change of the task: To avoid the uncertainty of what the task is, the task was elaborated more and it was tried to clarify what the students were supposed to do as well as explicitly trying to direct the focus towards the standard parabola by including a graph of the standard parabola as a *starting point*. The scaffolding was implemented directly underneath the corresponding task to make the *sequence of tasks* more explicit. This change in the task shifts the focus more on the connection between the symbolic and graphical representations and also tends to focus more on the Grundvorstellung object. A second task was added which specifically asked to determine reasons for the influence trying to reduce the *size of discoveries* regarding the explanations. Students testing the minimally guided approach tended to write each on their own paper and not discuss what their results were. To evoke more discussion about the results of the investigation, a third task was also added in which students were asked to write a cheat sheet with all their results including reasons. These summary sheets were then to be collected in order to analyse a greater amount of results by the students.

Conclusion for the Minimally Guided Approach
The testing of the minimally guided approach showed that it was not enough guidance to evoke the intended discoveries. This result is in line with the research that too open tasks while discovery learning do not foster the learning enough. The *size of discovery* (as described in section 4.2) has to be decreased in order to achieve more success. The first testing led to good hints for further development which resulted in the second version.

7.2.2   Second Version: an Approach Between Minimally Guided and Guided

The second version of the task first includes the graphs of $f(x) = x^2$ and $f(x) = 4 \cdot (x - 2)^2 + 3$ in contrary to the first version, which did not include any graphs (see Figure 7.13). The two graphs were included in the task in order to familiarise the students with a standard and a transformed parabola and to make the *starting point of the discovery* more clear. The intention was that the differences in the two graphs stimulated curiosity of the students and result in motivating a meaningful investigation. The students are then asked in a first task to investigate the changes of the graph for the general vertex equation and change the three parameters. To avoid the confusing action of changing all parameters at once, the hint is given to change only one parameter at a time as well as sketching the graphs in the hope of reducing the *size of discoveries* necessary. In this task a specific instruction for each of the groups is given in italics, in order to avoid confusion regarding the use of technology. In Fig. 7.13 instructions for all four groups are seen, the students were only given the one corresponding to their group. The *function plotter* instruction prompted students to look at different examples and plot the resulting graphs with the handheld calculators. For the *drag mode* group the drag mode was explained and it was requested to look at different examples for the parameters and the graphs of the resulting functions, while for the *sliders* group the hint focusses on changing the graph in the given file with the sliders. In the *without visualization* group task the attention was only drawn to looking at different examples and their corresponding graphs. For the full German version of the task see section B.2.3.

In this task, students would need to change between different representations to gain meaningful insights. In particular, students would need to choose numerical values for each of the parameters a, b and c, input them into the function equation, which would need some algebraic skills and then produce a graph of the resulting function with the means available (either per hand or one of the three technical visualizations). Afterwards, they would need to interpret the resulting graphs and connect the change in the graphical representation with a change in the symbolic representation. This task implements more scaffolding than the first open discovery version described in subsection 7.2.1 as students are directly asked to change the parameters one by one and therefore the necessary *size of discoveries* was intended to be more managable for the students.

Task 2 concerns the reasons behind the change. To support the finding of explanations again a hint is given. This time all groups are given the same hint, which was to look at the function tables. In the three experimental groups these could be displayed with the technology. The function tables can help with the reasons for the influence of the parameters, because they focus more on the Grundvorstellungen mapping and covariation than on the Grundvorstellung object. Therefore students were thought to be more likely to identify that shifting the parabola vertically is the result of adding a constant, while shifting it horizontally "moves" the x-axis when viewed with the rubber sheet metaphor described in section 3.4. Compared to the open discovery approach, the task directly asks for reasons as part of the task and not only in the instructions. The last task aims for the students to write their findings down by creating a cheat sheet including the reasons behind the transformations of parabolas. It was intended to prompt discussion between students while they were designing the cheat sheet and therefore facilitate reflection of their work. It was implemented after the open discovery approach to avoid the possibility that the students working in pairs mindlessly write something down on their own.

Testing of the Intermediate Guided Approach

The second version of the tasks was not tested with high school students, but piloted with about 100 first-year university students instead. Firstly, first-year university students often encounter the same difficulties with tasks than school students, secondly they can be utilized as an expert group to improve and evaluate the task from a didactical standpoint. The university students were from a pre-service teaching course for the lower secondary schools and in June 2015 taking a course on "Algebra and Functions in lower secondary school". The course consists of a two-hour lecture and a two-hour tutorial per week. In this course TI-Nspire CX CAS is used during the lecture and tutorial as well as being permitted in the final examination. The students had roughly 6 weeks of lectures before this topic, so they were able

You have learned about the function $f(x) = x^2$ and its graph. Here is the graph again as a reminder.

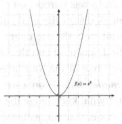

With factors and summands the function equation can be changed. For example the graph of the function $f(x) = 4 \cdot (x-2)^2 + 3$ looks like this:

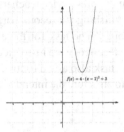

Task 1:
Work in pairs and try to find out, what changes occur in the graph of the function $f(x) = a \cdot (x-b)^2 + c$, when you change a, b and c. Maybe it helps you, if you only change a at first, then b and then c. Try sketching the graphs.

Without Visualization: *Look at different examples and the corresponding graphs.*
Function plotter: *Look at different examples and plot the functions with the calculators.*
Drag mode: *If you move the cursor to the graph and hold it, you can move the graph and stretch or shrink it. Look at different examples by those means and the corresponding graphs.*
Sliders: *Look at different examples and the corresponding graphs with the handheld by changing the function with the sliders.*

Task 2: Try to give reasons why a, b or c cause the corresponding change in the graph.
Hint: Look at the function table of the function and what effect a change of a, b or c has.

Task 3:
In pairs, create a kind of cheat sheet on a DIN A4 paper, where you describe with reasons, how the graph changes. The cheat sheet should be designed in a way that students from a different class understand it.

Figure 7.13 Second Version of the Tasks for the Intervention, Scaffolding for Each Group in Italics

to work with the TI-Nspire CX CAS system, thus an introduction was not needed. In the lecture the standard parabola was introduced and in the tutorials the task for the intervention was tested, but as one of the goals of the lecture is to prepare the pre-service teachers for their future job, the university students did not just solve the tasks. Instead, the jigsaw method was used and therefore expert groups were built first trying to solve the task with one of the four intended groups (*without visualization, function plotter, drag mode, sliders*) on a student level. Solving tasks on a student level taking into account what knowledge the students have is an important aspect of the pre-service teachers education, as teachers have to be able to determine if an intended task is manageable by their students. The pre-service teachers worked for 30 minutes in the expert groups. This was less than in the first version, however, it was anticipated that the pre-service teachers would be faster than school students. Also it was aimed at finishing the second part of the jigsaw method in the same tutorial as the expert groups. The task for the expert groups can be seen in Figure 7.14 and the pre-service teachers were instructed to work on the task in pairs solving it on a Year 9 level. The task also asked for their likes and dislikes and ways of improving the task in order to structure the intended discussion in the second part of the jigsaw method.

Task (same topic groups): Try solving the Year 9 task below in pairs. You will be sorted into four groups and solve the task with different ways. Like a Jigsaw, we would like to discuss the different ways with you. While working in the pairs, please think about what you like and dislike at the task and how the task could be improved.

Figure 7.14 Task for the Same Topic Groups

Then, as intended with the jigsaw method, discussion groups were conducted. To structure this discussion a reflection worksheet was given, which can be seen in Figure 7.15. In the discussion the students were asked to evaluate the potentials of the different groups (*without visualization, function plotter, drag mode* and *sliders*) and the task itself and at the end of the discussion choose their preferred way. This part was mostly implemented for the impressions of the pre-service teachers on the tasks. If all members of the discussion groups gave their permission the discussion was audiotaped and the reflection worksheets collected from those students. The staff conducting the tutorials and the researcher were also present during the tutorials and their impressions were collected where possible. The discussions were also used as a learning opportunity for the pre-service teachers regarding task design and the use of technology in inquiry situations.

Task (discussion groups): You have worked on the same task with different technical tools.
  a) Please tell your group about your way of solving the task.
  b) Discuss the following points in your group:
      • The task design
      • The used media and their differences.
  c) Please note your discussion points below and think about, which way you prefer.

	group 1 (without technology)	group 2 (function plotter)	group 3 („drag mode")	group 4 („slider bar")
General comments				
advantages				
disadvantages				

My preference:

Figure 7.15 Reflection Worksheet for University Students

Evaluation of the Second Version

The collected audiotaped discussions and the impressions of the staff and researcher were used to evaluate the task. The overall impression of testing the intermediately guided version was that the tasks were more comprehensible than the first version, however, still some difficulties remained. These difficulties and the implications for a change of task will be presented in the following.

One of the most common problems was that the 30 minutes planned for the task was not enough time for the students to finish it. This problem led to stress and due to this the students skipped parts or did not think their examples through.

Implications for a change of task: As a consequence, the time frame was extended for the intervention to 60–70 minutes.

One problem experienced and discussed by a number of students was that not enough or non-meaningful examples were looked at in Task 1 (see Figure 7.14) and therefore not all transformations were correctly discovered. Another problem was that students changed all parameters a, b, c at once and therefore could not differentiate between the different influences of each parameter. The hint to only change one parameter at the time was not explicit enough to induce all students to follow this.

Implications for a change of task: Even though the students testing the task were university students studying to become teachers, they did not discover all transformations and reasons for the ones found were sparse. This led to the assumption that for high school students even more structuring and guidance is needed to lead the students to discovering and explaining the influences. It should especially be some kind of structuring provided, which prevents learners from changing all parameters at once. So it follows that the two design principles *sequence of tasks* and *size of discoveries* (as described by MOSSTON 1972, section 4.2) have to be implemented and elaborated even further.

Another problem occurred with the phrasing of the writing task. Students were not sure what was meant by a cheat sheet and what they should write.

Implications for a change of task: The writing task needed to be rephrased to try to make it clearer what was expected from the students to include on the cheat sheet.

The necessary changes were implemented and a third, final version was developed. This version will be described in the next section 7.2.3.

7.2.3   Final Version: A Guided Discovery Approach

After testing the tasks in the tutorials to the university lecture a final version using an even more guided discovery approach was developed. It was therefore again more structured for the purpose of making the discovery more guided and achieving a successful learning (see also chapter 4). For this the *sequence of tasks* was made even more explicit and it was intended to reduce the necessary *size of discoveries*. The final version had four parts to guide the students through their discovery. The four parts (Describing an example, Investigating graphs, Finding Explanations, Creating Summary Sheet) will be described in the following including the changes compared to the second version and the aim of the re-designed tasks.

First Part: Describing an Example
As a *starting point of the discovery* the students were given the equation of a standard parabola $f(x) = x^2$ and a transformed parabola $f(x) = 4 \cdot (x - 2)^2 + 3$ and their graphs and were asked to describe the differences between the two graphs in the first part (see Figure 7.16) of the worksheet. This aspect was kept on from the intermediate guided version (subsection 7.2.2) as it was successful in arousing curiosity and giving the students a first feel for the topic of the investigation. Also this part represents a threshold into the task, which all students were thought to be capable of. The students did not need to interpret the two graphs in detail, but describe that the right graph is at a different position in the coordinate grid and appears smaller. They therefore first had to recognize the connection of the symbolic and graphic

First Part: Describing an example

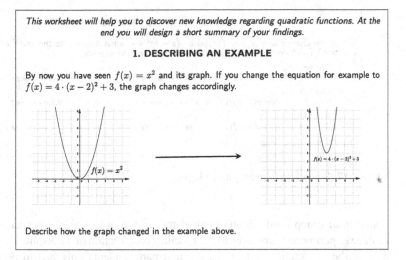

Figure 7.16 First Part: Describing an Example

representations and then had to compare two given graphical representations. In this part the students did not need to sketch the graphs themselves or use the technology provided.

Second Part: Investigating Graphs

In the second part (see Figure 7.17) of the worksheet the students investigated the influence of the different parameters a, b, c in the quadratic functions. To structure this task the students were offered a possible structuring by looking first at parameter c in $f(x) = x^2 + c$, then parameter a in $f(x) = a \cdot x^2$ and lastly at parameter b in $f(x) = (x - b)^2$ and by this providing a possible *sequence of tasks*. But this was just an offer, which was stated by the teacher. This structure was chosen with the aim of reducing the necessary *size of discoveries*, because the university students, who tested the second version (see section 7.2.2) tended to change more than one parameter at the same time and therefore did not discover the differences between the parameters. The parameters were split in the order c, then a and lastly b, because it was conjectured that the influence of c is the most accessible of the three, and the influence of b is the hardest to understand. This hierarchy in difficulty has been described in the literature (EISENBERG & DREYFUS 1994, GADOWSKY 2001, KIMANI 2008, see also subsection 2.2.2 and section 3.4). In this part of the intervention the students used the technology allowed in their group, so the *without*

Second Part: Investigating Graphs

---

2. INVESTIGATING GRAPHS

When you change a, b or c in $f(x) = a \cdot (x - b)^2 + c$, what happens to the graph of the function? The following steps might help you to answer the question.

1. First look at c.
    For c in $f(x) = x^2 + c$ use different values, for example -5, -2, -1, 0, 1, 2, 5. What changes then?

2. Now look at a.
    Use different values of a in $f(x) = a \cdot x^2$ for example -5, -1, $-\frac{1}{2}$, 0, $\frac{1}{2}$, 1, 5. What changes?

3. Finally look at b.
    Use different values of b in $f(x) = (x - b)^2$ for example -5, -2, -1, 0, 1, 2, 5. What do you see?

Use the media provided.

Figure 7.17 Second Part: Investigating Graphs

visualization group used scientific calculators and sketched graphs per hand and the three experimental groups used the specific configuration of TI-Nspire CX CAS (as described in section 6.4). As in the first part, students only had to describe the changes in the graph with the hope of producing a *Eureka-Moment*. However, depending on the group, different changes of representations were focussed. While in the *without visualization* group the change between symbolical via numerical to graphical representation was focussed, in the two dynamic groups it was more a treatment in the graphical representation and only a lesser focus on the change between the symbolic and graphical representations.

Third Part: Finding Explanations
In the third task (see Figure 7.18) the aim was to find explanations for the transformations. Again a structure to support was given through splitting the task and looking at each parameter individually. As with part two, this structuring was added after the trial with university students showed that finding explanations was not sufficiently achieved with less structuring and therefore it was intended to reduce the *size of discoveries*. The hints were designed to provide the students with a starting point of reasoning. Congruent to part two of the task, the structure began with parameter c by giving two conflicting statements of two fictitious students. The students were then asked to discuss with their neighbour, whose statement is correct. The statements were formulated with an intention to create a conflicting point of view in order to present a starting point for meaningful discussions. It was intended that the students realised that the parabola was only shifted vertically and it only seemed narrower, therefore overcoming the illusion described by GOLDENBERG (1988) (see subsec-

Third Part: Finding Explanations

---

3. FINDING EXPLANATIONS

Now try to find an explanation for your findings in exercise 2. The following steps may be helpful again:

1. Try to explain the influence of c.

> The graph
> gets narrower
>
> Sonja

> No, it stays
> the same!
>
> Cem

$$f2(x)=x^2+1$$

$$f1(x)=x^2$$

x	f1(x):= $\blacktriangledown$	f2(x):= $\blacktriangledown$
	x^2	x^2+1
-6.	36.	37.
-5.	25.	26.
-4.	16.	17.
-3.	9.	10.
-2.	4.	5.
-1.	1.	2.
0.	0.	1.
1.	1.	2.
2.	4.	5.
3.	9.	10.
4.	16.	17.
5.	25.	26.

Discuss with your neighbour if Sonja or Cem is right.

2. Now try to explain the influence of a.
   Look at your findings in exercise 2 again and compare to the graph of $f(x) = x^2$. Try to describe how the graph changes depending on the value of a.

3. Finally try to explain the influence of b.
   Sketch the graphs of $f1(x) = x^2$ und $f2(x) = (x-1)^2$ in one coordinate system and try to understand the following statements.

> If I look at (x-1)²
> instead of x² all values
> in the table move
> towards bigger x

> x² =9 is correct for x=3,
> but (x-1)²=9 first
> applies for x=4.

Do you have a different explanation?

Figure 7.18 Third Part: Finding Explanations

tion 2.2.3). One way to argue for the only vertical shift could be using the function tables also provided in the task and comparing the y-values of two functions for the same x-values. It could then be realised that the y-values constantly differ in only c and therefore the width of the parabola could not change. This explanation would focus heavily on the Grundvorstellung mapping.

For parameter a the students were directed to look at their findings in part two again and compare with the standard parabola $f(x) = x^2$ and then try to explain the graph changes depending on the value of a. Here, the intention was that students realised that the parabola was narrower than the standard parabola if a was greater than 1 or smaller than -1 and wider than the standard parabola if a was between -1 and 1. Also the students were supposed to realise that the opening of the parabola depended on the sign of a, namely the parabola is opening upwards if a is positive and downwards if a is negative. The explanations of this influence utilize more the Grundvorstellung function as object.

As it was expected that reasoning for parameter b was the most difficult, two correct explanations for the influence of b were given and the students were asked to first understand those and then look for another one. The two explanations focus on the Grundvorstellungen covariation and mapping, the left one focussing more on the covariation, the right one more on the mapping. Also the right explanation gives students a starting point for algebraic manipulations in order to achieve some insight.

Fourth Part: Creating Summary Sheet

Fourth Part: Creating Summary Sheet

4: CREATING SUMMARY SHEET

On a DIN A4 paper gather all your findings and present them in a way that someone else can understand.

Figure 7.19 Fourth Part: Creating Summary Sheet

In the last part (see Figure 7.19) the students summarized all of their findings. The students were asked to write a summary sheet to collect all explanations, findings etc. from the first three parts. The method of writing a cheat sheet was kept from the second version of the intervention task, however it was reformulated in order to make clearer what the students were supposed to write down. The students were asked to write all their insights, results and explanations down. It was explained verbally by the teacher that the students could imagine writing a cheat sheet for some friend who was allowed to take it into an exam on quadratic functions, but was not present in class during the intervention. Apart from being used as a source of data, the intention of the task was to induce reflection of the students work, which can lead to greater insight as well as clearing up ambiguous results.

The final version of the intervention task was then used in the main study with 14 classes of Year 9.

Due to the study being conducted in Germany in German speaking classes all analysis, developing of the coding manuals and coding was done with the original German materials. Presented in the text here are the translated versions by the author. It was tried to translate as accurate as possible. The summary sheets were coded using qualitative content analysis according to MAYRING (2004). The coding manual was developed using the summary sheets of two classes with the help of two master thesis students TERFURTH (2016) and NABERS (2016) and one state examination[1] student KLEIN ALTSTEDDE (2016) closely monitored by the researcher.

8.1 Qualitative Content Analysis

The method used for analysis of the data was *qualitative content analysis* (QCA) as described by MAYRING (2004) (see also MAYRING 2015; MAYRING 2016). The method and steps taken will be described in this section. QCA was developed in order to transfer the systematic analysis used by content analysis while still taking a qualitative analysis point of view. One follows a number of steps guiding the qualitative analysis using theory as a guideline, but also the structure and content of the material that is to be analysed. It can be differed into three different types of QCA *summarizing, explicating* and *structuring* content analysis (MAYRING 2004). For this thesis a mix of the summarizing and the structural approach was chosen, in which particular aspects are selected from the material and structured using in part from theory derived categories. This adheres to the structural approach. A

---

[1]All students studying to become teachers and beginning before 2011 at the University of Duisburg-Essen obtained a state examination degree. After 2011 it was changed to a master's degree.

© The Author(s), under exclusive license to Springer Fachmedien Wiesbaden GmbH,   109
part of Springer Nature 2021
L. Göbel, *Technology-Assisted Guided Discovery to Support Learning*,
Essener Beiträge zur Mathematikdidaktik,
https://doi.org/10.1007/978-3-658-32637-1_8

vast number of categories was then developed inductively from the material. The concrete steps in developing the coding manuals for the different aspects of analysis will be described in more detail in the respective sections.

8.2   Analysis of Pretest Results

The pretests were scored and points for each item were input into SPSS statistics separately. Each pretest score was connected to the students' code which the students generated themselves. The points distribution as well as cross-tabulations for frequency across the groups were then produced by SPSS. All statistical analysis was conducted using SPSS.

8.3   Analysis of Summary Sheets

Using the method described in section 8.1 two coding manuals for the summary sheets were developed in a number of steps.

The first coding manual was developed with the help of two master thesis students TERFURTH (2016) and NABERS (2016) and one state examination thesis student KLEIN ALTSTEDDE (2016) and was aimed at a comprehensive analysis of the collected summary sheets (subsection 8.3.1). The comprehensive analysis regarding the summary sheets aimed at giving an overview about what students found out in the intervention as well as describing the summary sheets in different categories. In a first step four main categories were set normatively in order to structure the development. The four categories were *structure of the summary sheet, use of representations, language use,* and *content*. Use of representations and content were chosen as they can be used to determine the students' understanding regarding the influence of parameters. Also the language use, so what kinds of technical terms the students use, points to the students' sophistication in regard to the knowledge. The structure analysis of the summary sheet was included to review the task wording as well as keeping open the possibility to conduct a statistical analysis if for example students who design a structured worksheet have greater knowledge.

The second coding manual was implemented after the comprehensive analysis due to interesting results and aimed at a more in-depth analysis of answers regarding parameter c. This in-depth analysis was conducted with the help of a master thesis student KURTUL (2018) (subsection 8.3.2).

8.3.1 Structure, Representation, Language Use and Content

Taking the four main categories as normatively set, sub-categories were developed from the material using the summary sheets of two classes. For a broad overview of the resulting summary sheets a class from the *without visualization* group and a class from the *sliders* group were chosen. Subcategories were developed from this material inductively and then discussed with the researchers' group and the student coders.

In total 84 subcategories were developed. The majority of the categories, in total 69 of 84, had three specifications, namely *not written, yes, viable*, and *yes, non-viable*[2]. The full coding manual can be found in Appendix C.

All classes were then coded by at least two raters (the author and one or more student coders). The codings were recorded in the statistics software SPSS and analysed as described in section 11.1. The double coding was implemented in order to determine interrater reliability. It showed a middle to high accordance in the coding (classification of the coding using cohen's kappa as described in DÖRING & BORTZ 2016, p. 346). If more than two raters coded a category cohen's kappa was computed for all pairs and the median will be given here (as described in DÖRING & BORTZ 2016, p. 346). In the categories regarding parameter b and c there were some adjustments to the coding manual after coding by the student coders, thus it was then independently coded again by a student helper and the interrater reliability also tested against her. For two of these categories (namely horizontal and vertical influence) the interrater was then still very low, so a third rater, a fellow PhD-student of the author coded all summary sheets, where the previous two raters did not agree and through this a majority decision was reached for these students. The interrater was then computed against the majority decision.

8.3.2 Students' Reasoning

While coding the summary sheets using the coding manual described above, the coders realised that the content-related codes (see subsection 8.3.1) were rather superficial only differing between viable and non-viable answers regarding the influence. It was then decided to take a closer look at the answers on the summary sheets. The answers regarding parameter c were chosen for developing a coding manual. An expert group consisting of the author, her PhD-supervisor and doctoral colleagues

---

[2]The German original were "Nein, ja tragfähig, nicht tragfähig". Viable was chosen as the best translation.

inductively developed a categorization of the answers regarding parameter c taking one control-group and one sliders group class into account. It was done in pairs, but the categories were clearly seperated and conformed over the five different pairs of experts. The coding had two dimensions (see Table 8.1). Firstly all answers could be clustered into static or dynamic answers, mainly based on the wording the students chose. The answers also fell into one of five topic-related categories: *y-intercept, y-axis, parabola shape, vertex point, slope*. For each category it was also decided if it was *viable, partly viable* or *non-viable*.

Table 8.1 Categories Developed for the In-depth Analysis

	y-intercept	y-axis	parabola shape	vertex point	slope
static					
dynamic					

The summary sheet answers regarding parameter c were then coded by the author and a master thesis candidate (KURTUL 2018). The categories y-axis and vertex point were differentiated into sub-categories. However the numbers of answers were then rather small, so no further analysis of the sub-categories were undertaken and the sub-categories not presented here. Interrater reliabilty was middle to high (Cohens Kappa of 0.68, with two codings identified as matching if 85% was the same text segment) so no further adjustment of the coding manual was done.

8.3.3   Quantitative Analysis of Summary Sheets Results

For the comprehensive analysis of the students' summary sheets the categories were also input into SPSS statistics and all coding recorded there. Cross-tabulation and chi-square tests were then conducted through the software as well as interrater reliability computed.

8.4   Analysis of Videos

In order to better understand the learning pathways and processes when solving the tasks, 13 pairs of students were videographed during the intervention. The videos were then used as a source for case studies. The literature review provided some guidance for the selection of these case studies through presenting interesting aspects to be searched for. The uses of technology in experimental mathematics (as described by BORWEIN 2005) seemed promising. Also theoretical frameworks such as the communication framework (developed by BALL & BARZEL 2018, see section 3.5) were used to identify sequences in the videos, which all show a potential of the technology for the students learning. Apart from the use of technology and communication it was focussed on the concept of parameters and the evaluation of the designed sequence of tasks. Possible misconceptions and obstacles regarding these two aspects might provide valuable clues for a further development of the Guided Discovery approach.

A number of other interesting sequences could be identified in the videos, which could be classified into potentials and constraints of the approach and linked to the affordances and constraints described in the literature (see section 3.2).

To first identify the interesting sequences regarding the above mentioned aspects and following the first steps in a summarizing content analysis (MAYRING 2004) all 13 videos were transcribed and so called solution summaries were written. The solution summaries are shorter descriptions of the videos, which describes the video in shorter phases, the phases were for example the work on one subtask of the worksheet or if the students clearly changed the topic of their investigation. The solution summaries also included the investigated functions and the use of the technology. This phasing was supported through three masters-thesis students WESSELER (2016), HÖNEKE (2019), MOSKALENKO (2019) and the phasing discussed within the researchers' group. The phasing enabled an identification of common patterns in the learning pathways across the 13 videos.

To gain further inside into the structure of the processes, the processes visible in two videos were chosen to be visualized using timelines including the graphed functions, the use of technology and important aspects of the Guided Discovery. This short graphic representation can lead to identification of common uses of technology in the processes (see section 12.1 for the results of this graphic timeline analysis).

After the two examples of Guided Discovery that were analysed through graphical timelines, aspects regarding potentials (see subsection 12.2.1) as well as constraints (see subsection 12.3.1) were identified. Also in the intervention students were asked to explain the influences, this task led to both early versions of explanations (subsection 12.2.4) as well as the problem that no explanations were evoked (subsection 12.3.4). Content-related constraints were identified in the case of parameter stereotypes (subsection 12.3.3) and the value of a in the standard parabola (subsection 12.3.2). All chosen sequences were discussed with the researcher's groups and analysed as case studies.

Part IV
Results

Data

9

The researcher was able to conduct the study described in Chapter 6 in a total of 14 classes during the 2015–2016 school year. In this chapter an overview of the participating classes and the data that was collected during the study will be given. Twelve teachers taught the classes across seven schools. If the teachers did not feel comfortable to conduct the intervention themselves, the researcher supported them and conducted the intervention. This was the case in 4 classes.

The necesssary information about participating classes (section 9.1) and an overview over the collected data (section 9.2) will be given here.

9.1 Participants

A total of fourteen classes took part in the intervention. The classes were from seven schools in North-Rhine Westphalia and one school in Thuringia and were taught by twelve different teachers and acquired through personal contacts of the researcher and her working group as well as an appeal in a newsletter for Teachers Teaching with Technology, a network of teachers. Two teachers taught two classes each. In 13 of the 14 classes (all in North-Rhine Westphalia) the researcher was present during the intervention and supported the teacher if wished. In four of these classes the researcher conducted the intervention herself. The schools are all upper secondary schools. Three of the schools are from one city, two from the same district. One school was willing to participate with a total of three classes. The classes used different calculators in the course of year nine. Only one class already used the TI-Nspire CX CAS, while 6 classes used the scientific calculator Casio fx-991 before the intervention. Seven classes already used graphics calculators though different models (two classes used TI82, three classes used TI84+C, two classes Casio CG20).

© The Author(s), under exclusive license to Springer Fachmedien Wiesbaden GmbH, part of Springer Nature 2021
L. Göbel, *Technology-Assisted Guided Discovery to Support Learning*,
Essener Beiträge zur Mathematikdidaktik,
https://doi.org/10.1007/978-3-658-32637-1_9

For the intervention it was decided that if more than one class was taught by one teacher or were from one school they would be assigned different groups. The classes were identified through a teachers code as well as the experimental group. The students participating were also assigned a code, which they used both in the pretest and the intervention. As described in section 6.1 the experimental groups all used TI-Nspire CX CAS preferably on iPads, but in case the iPads were not available TI-Nspire CX CAS handhelds were used. In the intervention one class in the *drag mode* and one class in the *sliders* group used TI-Nspire CX CAS handhelds. All other 9 classes in the three experimental groups used TI-Nspire App on iPads. For the *without visualization* group scientific calculators were used, except in one class from the *without visualization* group, where students did not possess scientific calculators without graphing facilities and therefore used no scientific calculators. The other two control-group classes used Casio fx-991 in the intervention. For a comprehensive overview of the school number, the used calculator in the intervention and as well as the teacher code see Table 9.1.

Table 9.1 Overview Over the Classes

group	teacher	school number	used calculator	Intervention with
Without visualization group	ALE26	1	fx-991	Scientific calculator
Without visualization group	GHR02	2	fx-991	Scientific calculator
Without visualization group	SAR02	3	TI-82+	no calculator
Function plotter	GIA22	4	TI-84+C	iPads
Function plotter	NOA08	5	fx-991	iPads
Function plotter	ROO12	6	CG20	iPads
Drag mode	LHR17	5	fx-991	iPads
Drag mode	ROT14	4	TI-84+C	iPads
Drag mode	RYO13	2	fx-991	Handhelds
Sliders	NAM03	7	TI-Nspire	Handhelds
Sliders	RAU01	8	fx-991	iPads
Sliders	ROO12	6	CG20	iPads
Sliders	SAR02	3	TI-82+	iPads
Sliders	WLA31	4	TI-84+C	iPads

9.2   Data Overview

During the study different data were collected (see Chapter 6) i.e. the pretests to collect baseline data and summary sheets during the intervention (see subsection 7.2.3). Where it was possible a pair of students was videographed during the intervention, while working on the series of tasks. This was achieved in thirteen of the fourteen classes. For these classes, the videos and their transcripts were part of the collected data. In 13 classes the researcher was present and for these classes there are lessons observation notes. 383 students took part in the study. However, due to collecting data while flu season not all students took part in all parts of the study. So 357 students wrote the pretest, also 357 students were attending the intervention. These are not necessarily the same students, as some students were not in class for the pretest, but for the intervention and vice versa. For 310 students, the summary sheet could be connected to a pretest result. 26 of the 357 students attending the intervention were videographed in pairs. These students were told that they did not necessarily had to produce a summary sheet during the intervention in order to encourage them

Table 9.2 Overview Over the Collected Data

Collected Data	Without visualization	Function plotter	Drag mode	Sliders	Total
Number of participating students	81	77	91	134	383
Number of collected pretests	78	70	89	120	357
Number of collected summary sheets	35	33	44	66	178
Number of students who produced summary sheets	71	67	85	130	353
Number of videographed students	6	6	6	8	26

to discuss more and not focus on the production of the summary sheet. However, twenty-two of the twenty-six produced one. All other students in the intervention were asked to produce a summary sheet and did so in groups of two or three. 178 summary sheets were collected in total. For the analysis the summary sheets were assigned to each of the students, who produced it. So if two students, A and B, designed the summary sheet, all coding was once counted for student A and once for student B. Table 9.2 provides an overview of the distribution of students numbers, number of pretests and summary sheets according to their group, while Table 9.3 shows the distribution split into classes.

Table 9.3 Number of Students' Collected Data Divided by Classes

Group	Teacher	No. of collected summary sheets	No. of students	No. of students pretest	No. of students summary sheets
Without visualization	ALE26	10	26	26	20
Without visualization	GHR02	14	31	29	28
Without visualization	SAR02	11	24	23	23
Function plotter	GIA22	11	25	25	25
Function plotter	NOA08	12	25	25	22
Function plotter	ROO12	10	27	20	20
Drag mode	LHR17	14	32	32	28
Drag mode	ROT14	14	29	27	29
Drag mode	RYO13	16	30	30	28
Sliders	NAM03	13	26	22	26
Sliders	RAU01	12	24	22	24
Sliders	ROO12	14	27	27	26
Sliders	SAR02	14	29	24	28
Sliders	WLA31	13	28	25	26
Total		178	383	357	353

Pretests

The pretests were corrected and scored and then the scores analysed using SPSS statistics. Pretests only served as a baseline in order to determine if the intervention groups were comparable. For this purpose the maths grade of Year 8 was also collected.[1] A total of 56 points could be achieved in the pretest. Even though the topics of the pretests should have been known by all students, the test proved to be difficult for some students. The lowest overall score was 13 points, the highest 54 points with a mean of 35.34 (std.-deviation 7.54). While the overall scores show only a general impression of the test (section 10.1), some items show interesting results also by the chosen way of solving the items by the students (section 10.2). However, not all items described in section 7.1 were in retrospect relevant for the intervention, they still provide certain insight to form a consistent view of the knowledge of the students.

10.1 Mean Test Scores

Mean test scores of the reached points in the test (max. 56 points) and the potential differences between intervention groups or classes[2] were analysed statistically.

The differences between the mean test scores in Table 10.1 of the different experimental and control groups show a slight tendency towards the *without visualization* group performing better on the pretest. Analysis of variance showed that these dif-

---

[1] German maths grades are 1 through 6 with 1 being the best and 6 being the worst and would lead to necessity of revising a school year.

[2] Classes are identified by the intervention group number 1 to 4, with one being without visualization, 2 function plotter, 3 drag mode and 4 sliders and the teacher code as stated in Chapter 9.

© The Author(s), under exclusive license to Springer Fachmedien Wiesbaden GmbH, part of Springer Nature 2021
L. Göbel, *Technology-Assisted Guided Discovery to Support Learning*,
Essener Beiträge zur Mathematikdidaktik,
https://doi.org/10.1007/978-3-658-32637-1_10

Table 10.1 Pretest Scores in the Different Intervention Groups

Group	N	Mean pretest score	Standard deviation	Median
Without Visualization	78	37.173	7.431	37.00
Function plotter	70	35.436	7.035	35.50
Drag mode	89	35.034	6.040	36.00
Sliders	120	34.321	8.690	34.00
All students	357	35.34	7.544	35.50

Table 10.2 Mean and Median of the Pretest Scores in the Different Classes

Group	Teacher	N	Mean test score	Standard deviation	Median
Without Visualization	ALE26	26	37.788	9.090	36.00
Without Visualization	GHR02	29	38.845	5.323	39.00
Without Visualization	SAR02	23	34.370	7.140	32.50
Function plotter	GIA22	25	37.440	6.261	38.00
Function plotter	NOA08	25	36.240	6.044	36.00
Function plotter	ROO12	20	31.925	8.059	32.25
Drag mode	LHR17	32	36.125	5.276	36.50
Drag mode	ROT14	27	31.667	5.549	31.50
Drag mode	RYO13	30	36.900	6.154	37.25
Sliders	RAU01	22	37.50	7.588	37.25
Sliders	WLA31	25	31.50	9.969	32.00
Sliders	SAR02	24	35.604	6.369	34.75
Sliders	ROO12	27	30.111	7.813	28.50
Sliders	NAM03	22	38.114	8.775	39.00

ferences are not significant on the 0.05 level, so it can be reasoned regarding the knowledge before the intervention the four groups are comparable. All intervention groups were assigned by class and it could be argued that mathematical knowledge is class or teacher dependent. Therefore, the mean scores of the pretests were also analysed by class in order to determine if there were classes that outperformed others.

In Table 10.2 the mean scores, standard deviation and median of each participating class can be seen. The highest mean score was achieved by the class 1GHR02 of the *without visualization* group, the lowest mean score by class 4ROO12 from the *sliders* group. Analysis of variances shows that the differences between the classes are statistically significant (p < 0.001). But multiple comparisons using Bonferroni correction show that this is due to the difference between the aforementioned classes, as only the difference between 1GHR02 and 4ROO12 is statistically significant on the 0.01 level. It was then decided to look at the maths grades of Year 8, which were also collected to see if there were also differences between the classes, however this

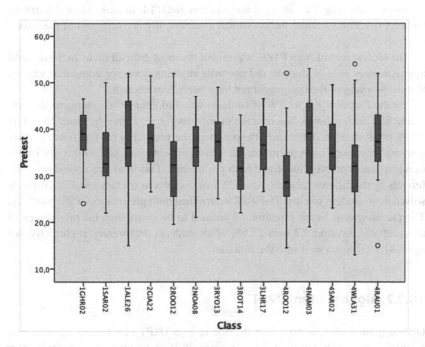

Figure 10.1 Boxplots of Pretest Results by Class

was not the case. Finally, the two extreme classes were looked at item for item and class GHR02 vastly outperformed the class 4ROO12 in the items D4JG and N1FQ which were deemed not relevant or badly posed for the aim of the study. Thus, it was decided to keep the two extreme classes in the study. Boxplots of the pretest results split by classes can be seen in Figure 10.1.

10.2 Pretest Results by Item

10.2.1 Finding Coordinates (Item P1ZC)

The notation of coordinates in item P1ZC (see Figure 7.1) was a relatively easy task with 75.6% of all students noting the correct coordinates of all four points. One observed difficulty was that the convention of the order of coordinates was obviously not correctly remembered as all coordinates were noted in the inverted order (so instead of (x, y) it was noted (y, x)). This was especially the case in the two classes taught by ROO12 and in class 3ROT14. In class 4ROO12 sixteen students (so 59.6%) noted the coordinates numerically correctly but in the wrong order.

The second part of item P1ZC was one of the most difficult items in the test and approximately 45% of students did not write anything down regarding it. It seemed that the wording of the task might not have been clear enough.

For the function $f1$, 41.7% of students who did attempt to determine the pre-image of 1 instead made the mistake of inputting $x = 1$ into the function, while 31.4% of all students determined the pre-image correctly. For the function $f2$ many students only found one pre-image, not the two pre-images 1 has, while only 14.9% of all students correctly identified both pre-images. This item was solved very differently by the different classes, while 35% of students from class 1ALE26 correctly solved it, no student of class 1GHR02 determined both pre-images of $f2$ correctly. The pre-images of linear function $f3$ seemed to be easier than the pre-images of the quadratic function $f2$ with 27.6% of all students determining it correctly, but again 41.7% input $x = 1$ into the function.

10.2.2 Algebra (Item U2PT)

Operating on term structures like the one in item U2PT (see Figure 7.2) also was relatively easy, with 249 of 327 students, so 76.1% who attempted the item solving it correctly. Incorrect solutions were mostly due to a computational error. The high quote of students solving it correctly suggests a ceiling effect of the item. Between the different intervention groups no vast differences in the solution rate were observed. Operating with term structures should only be needed to a small degree during the intervention for the three experimental groups working with technology, however, it might be needed by the *without visualization* group to simplify equations to graph functions.

10.2.3   Variable as an Unknown (Item E4FV)

Solving simple equations like the one in item E4FV (see Figure 7.3) $6 \cdot a + 3 = 45$ posed no significant problem to all students and thus a ceiling effect was observed as well. Only two students did not attempt to solve it and 94.4% of the rest solved it correctly. In some classes, namely 1ALE26 and 3RYO13 all students solved the equation correctly. The lowest solving rate was achieved in 4ROO12 with 80.8% solving it correctly. As the students in the experimental groups had computer algebra systems in the intervention, the slight disadvantage in solving equations should not be relevant for the results in the intervention.

In the second part of the item regarding the explanations of the way of solving, most students used some version of term transformations as explanation. About a quarter of the students stated that they used a multiplication table or just tried different values and did not use systematic term transformations. This way of explaining was strongly dependent on the classes the students belonged to, with 50% of 2ROO12 and only 8.3% of 4SAR02 used this. The difference probably stems from the tradition used and taught in the class, but as both ways lead to the correct solution they are both appropriate.

10.2.4   Fill Graphs (Item N1FQ)

Fill graphs (see Figure 7.4) were also quite easy for the students with only three students not attempting it and 76.6% of all students correctly choosing the fill graph that represented the filling of the swimming pool. If this was due to the graph as picture mistake as described in subsection 7.4 or because they understood it correctly, cannot be determined. In the class 4NAM03 all students solved the item correctly, while in 3ROT14 only 59.3% solved it correctly, the differences might be due to more familiarity with fill graphs in some classes than others.

10.2.5 Dependance of x and y (Item D3JG)

The covariation item (see Figure 7.5) was very much dependent on the class, if the students solved it correctly. While students in the class 1GHR02 only solved it correctly in 21.4% of the cases, 75% of the students in class 2NOA08 solved it correctly. One of the reasons for these vast differences could be that the covariation of linear functions, which are not in the form $y = m \cdot x + b$, is not so familiar for the students. This could be caused by the stereotyping of linear functions in the aforementioned form. Thus, some teachers might have taught the covariation of non-standard form linear functions and this lead to the better performance of the students.

10.2.6 Variable Concept (Item D4JG)

This item (see Figure 7.6) was ill-posed and led to various non-classifiable answers that showed students did not understand what was asked in the item. Points were awarded if the answers made some sense. For the second part of the item, one point was awarded for each meaning of the letters stated. 51.7% of students in class 1GHR02 stated meanings for all three letters, while only 11.1% of the students in class 3ROT14 stated meanings for all three letters. A further analysis of the answers was not conducted due to the wide range of different answers.

10.2.7 Change of Representation (Item A6XY)

As change of representation is a crucial part of understanding, the item A6XY (see Figure 7.7) was used to determine if students were able to connect between the different representations. The change between the graphical and tabular representation of two linear functions and $f(x) = x^2$ was comparable easy, only six students did not attempt it and 82.9% of students who attempted it connected them correclty. The results were however dependent on the class to some degree as all students in class 2GIA22 solved it correctly while only 65.2% in class WLA31 and 65.4% in class 4ROO12 solved it correctly. Connecting the tabular to the symbolical representation was attempted by less students that the connection between graphical and tabular representations (47 did not attempt it), but solved correctly by 83.9% of the students who attempted it. Calculating the missing values in the table was not attempted by about a quarter of the students. 97 students did not write anything for the missing values of the quadratic functions and 85 students did not write anything down for the missing values of the two linear functions. It might be reasoned that the relatively high rate of students who did not try to compute any values is due to the students not reading carefully and realising that there were empty spaces in the tables. 188 students calculated all six missing values correctly.

10.2.8 Family of Functions (Item Y3GJ)

Identifying a function equation for a family of functions (see Figure 7.8) was rather difficult, however, the different intervention groups differed in the solving rate. While 74.3% of the students in the *without visualization* group identified the function equation correctly and 67.2% of the students in the *function plotter* group only 50% of *drag mode* group students and 52.4% of the *slider* group students managed to do so. The differences were not due to one class, but all classes in the *sliders* group performed worse than all classes in the *without visualization* groups. One reason for the differences in the solving rate, might be unfamiliarity with family of functions. In order to correctly identify the equation one must identify both the slope and the y-intercept and those were the two aspects students mostly used to explain their choice. But the explanation was attempted by less students than identifying the equation with only 249 students attempting to explain. 40 students stated that they guessed the answer, this is different depending on the class, so for example in class 1ALE26 10 students (40%) and in class 3RYO13 9 students (34.6%) stated that they guessed, whereas in class 1SAR02 or 2GIA22 no student stated that they guessed.

10.2.9 Properties of Linear and Quadratic Functions (Items R5TG and R4TG)

Items R5TG (see Figure 7.9) and R4TG (see Figure 7.10) were very similar only differing in the degree of the function and some small aspects, which also shows in the results. But it was obvious that technical terms like first angle bisector were only taught in some classes, as some classes performed very well on these items while others performed very badly. This is understandable, if it is assumed that not all students were taught this technical term. Some other items, e.g. the one regarding b being the y-intercept were rather easy with some classes answering this 100% correctly. Again some classes performed much worse, e.g. only 50% of 2ROO12 answered this correctly. The reason for this might be that the class had been taught the y-intercept with a different letter. In both items the rate of students who decided for all 5 respectively 6 statements the validity correctly is very small. The quadratic function item R4TG only 4.2% of all students scored 5 points, while on the linear function item R5TG only 1.7% of all students scored 6 points.

10.3   Conclusion Pretest Results

Taking all items and the mean test scores into account, the pretest results show that even though some parts of the knowledge are teacher or class dependent these are not favouring one particular class enormously and thus the four intervention groups can be regarded as having equal potentials for the intervention.

Summary Sheets

In the intervention the students were guided to the exploration of parameters in quadratic function by a four part worksheet with the intent stated in the introduction to design a summary sheet at the end of the worksheet (see also subsection 7.2.3):

1. First students were asked to describe differences between the standard parabola and a transformed parabola.
2. Then students were asked to identify the influence of the parameters in the order c, b and then a in $f(x) = a \cdot (x - b)^2 + c$ by looking at the graphs.
3. Part three was intended to evoke explanations and also using the order of part two, the students were presented with tips, statements or even examples of explanations for the influence of parameters intended to lead students to new explanations.
4. Finally, in part four students were asked to design a summary sheet with all their findings and explanations. For this they were given a blank paper and no restrictions regarding the designing of the summary sheet given. This led to a variety of different designs with different quality in design, structure and correctness. Impressions of summary sheets produced are given in Figure 11.1.

Presenting the results of analysis of the summary sheets is the core of this chapter, while the analysis of the processes that took place during the intervention and were videographed will be presented in the next Chapter 12.

The results show differences between the dynamic groups (students using the drag mode or sliders) and the static groups (students using function plotters or without technological visualization) in the different categories regarding structure, language use, viability as well as detailed aspects regarding all parameters. The dynamic groups seem to be more suitable for the identification of parameter influence, however, part three of the worksheet does not succeed in evoking explanations

© The Author(s), under exclusive license to Springer Fachmedien Wiesbaden GmbH, part of Springer Nature 2021
L. Göbel, *Technology-Assisted Guided Discovery to Support Learning*,
Essener Beiträge zur Mathematikdidaktik,
https://doi.org/10.1007/978-3-658-32637-1_11

regardless of the intervention group. But with all results presented in this chapter it has to be taken into account that only the products were taken into consideration, so students of the groups that were not videographed might have found out considerably more during the intervention, which was then not noted on their products.

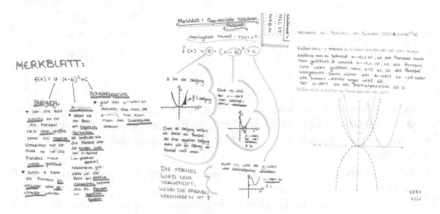

Figure 11.1 Impressions of Summary Sheets

11.1   General Results

As described in Chapter 9 178 summary sheets of 353 students were collected in the 14 classes. All summary sheets were digitalized and then coded in 84 categories (see subsection 8.3.1 for information on the coding manual). In this section the results of the quantitative analysis of this coding will be represented by referring to examples of statements on summary sheets and the statistical analysis of the coding presented and discussed.[1]

11.1.1 Structure of Summary Sheets

The overall structure show two main types of summary sheets that can be seen as representative for the depth of students' conceptualizations. Either students designed

---

[1] All summary sheets are in German, so all examples were translated by the author. Original German examples can be found in the Appendix D.

a summary sheet revolving around the vertex form of quadratic functions choosing to restructure all their during the process gained knowledge (*restructured*, for
an example see left side of Figure 11.2) or the students wrote short statements to
answer the corresponding task on the worksheet (*answers to task*, e.g. right side
of Figure 11.2). It can be conjectured that students writing a restructured summary
sheet gained deeper insight into the influences in their investigation and thus were
able to generalize from their answers to the task to a general knowledge summary
sheet. The *answers to tasks* kind of summary sheet follows more the given structure
of the worksheet. It could be interpreted that students who design those summary
sheets can argue their results less flexible and, therefore, only wrote answer statements to the given tasks.

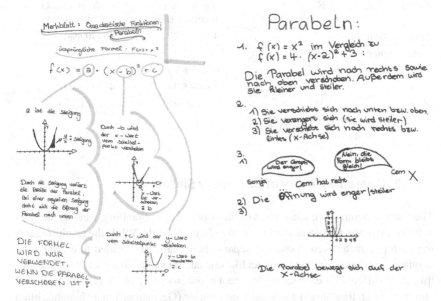

Figure 11.2 The Two Different Structures of Summary Sheets, Left *Restructured*, Right
Answers to Task

   Overall 65.7% of all students designed a summary sheet of the first category,
while 34.3% wrote answer statements. When one looks closer into the data, it can
be seen that it seems to not depend on the group the students are in Table 11.1. In
the *sliders* group 71.8% of students prepare a summary sheet of the first category,
while in the *function plotter* group only 59.8% fell into the first category. A chi-

square test showed that this difference is not statistically significant and it would need to be investigated further if the kind of technology used influenced the kind of summary sheet produced. A conjecture is that because students in the *drag mode* and *sliders* group had prepared files they did not write something down while they were working on the task. This might have led to a regrouping of knowledge once they reached part 4 and designed their summary sheet. On the other hand, students in the *without visualization* and the *function plotter* groups might have been more likely started writing down while working and handing these in as their summary sheets.

Table 11.1 Kind of Summary Sheet Designed

Group	N	restructured summary sheet	answers to tasks summary sheet
Without Visualization	71	42 (59.2%)	29 (40.8%)
Function plotter	67	40 (59.7%)	27 (40.3%)
Drag mode	85	61 (71.8%)	24 (28.2%)
Sliders	130	89 (68.5%)	41 (31.5%)
Total	353	232 (65.7%)	121 (34.3%)

p=0.236, Cohen's Kappa $\kappa = 0,852$

11.1.2 Overall Viability of Summary Sheets

The outer structure gave small hints to the depth of understanding, a closer analysis at the content was also conducted to provide further insight. While the results regarding each parameter were looked at separately, it was also decided to evaluate the summary sheet as a complete product like one might evaluate it to award a grade. For this it was differed for each summary sheet between *mostly viable* (German original "hauptsächlich tragfähig") and *mostly non-viable* (German original "hauptsächlich nicht tragfähig"). Examples for a mostly viable and a mostly non-viable summary sheet can be found in Figures 11.3 and 11.4. If the summary sheet contained both viable and non-viable statements the coder would balance all statements and decide which outweighed the others.

For example, the summary sheet in Figure 11.3 contains a hard to understand and maybe non-viable statement regarding parameter *b* ("If b gets bigger, the x-value also gets bigger"), but is still classified as mostly viable as the statements regarding the parameters *c* and *a* are viable. Compared to this, the summary sheet

Merkblatt mit Erkenntnissen, Ergebnissen und Begründungen

Summary Sheet with insights, results and explanations

Begründung zu C:
C verändert den y-Achsenabschnitt.
Wird die Variable c verändert, verändert sich auch der y-Achsenabschnitt, da die Variable x² formt den Graphen und c hat keinen Einfluss auf die Form.

Explanation for c:
C influences the y-intercept
If the variable c is change, the y-intercept
changes, because the variable x^2 forms the graph
and c doesn't have influence on the form.

Begründung zu a:
a verändert die Steigung.
Wird die Variable a verändert, verändert sich auch die Steigung, wird a höher, wird der Graph enger.

Explanation for a:
a changes the slope.
If the variable a is changed, the slope also
changes, if a is higher, the graph is narrower.

Begründung zu b:
b verändert den x-Wert.
Wird b größer, wird auch der x-Wert größer

Explanation for b:
b changes the x-value.
If b gets bigger, the x-value also gets bigger.

Figure 11.3 Example of a Summary Sheet With Mostly Viable Statements

in Figure 11.4 contains mostly non-viable statements. For parameter c the students stated the misconception of a changing width on the summary sheet, while with the statement for b is incomprehensible.

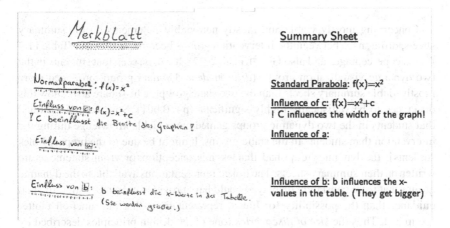

Merkblatt

Summary Sheet

Normalparabel: $f(x)=x^2$

Standard Parabola: $f(x)=x^2$

Einfluss von c: $f(x)=x^2+c$
? C beeinflusst die Breite des Graphen ?

Influence of c: $f(x)=x^2+c$
! C influences the width of the graph!

Einfluss von a:

Influence of a:

Einfluss von b:
b beeinflusst die x-Werte in der Tabelle.
(Sie werden größer.)

Influence of b: b influences the x-
values in the table. (They get bigger)

Figure 11.4 Example of a Summary Sheet With Mostly Non-viable Statements

The interrater for this category showed that there was a middle to high accordance in the coding[2], which is understandable as the balancing differs by each coder (Cohen's Kappa $\kappa = 0.589$, $\kappa = 0.632$, $\kappa = 0.781$ depending on the interrater). The results show that over three-quarters (77.3%) of all students produced summary sheets which contained mostly viable statements. This leads to the conclusion that most students did achieve some level of conceptualization of parameters in the intervention that allowed them to write mostly viable summary sheets.

Table 11.2 Overall Viability of Summary Sheets

Group	N	mostly viable statements	mostly non-viable statements
Without Visualization	71	45 (63.4%)	26 (36.6%)
Function plotter	67	45 (67.2%)	22 (32.8%)
Drag mode	85	75 (88.2%)	10 (11.8%)
Sliders	130	108 (83.1%)	22 (16.9%)
Total	353	273 (77.3%)	80 (22.7%)

$p < 0.001$, Cohen's Kappa $\kappa = 0.606$

Concerning mostly viable and mostly non-viable statements on the summary sheets, differences between the intervention groups become visible (see Table 11.2 for exact percentages and also GÖBEL et al. 2017). It can be seen that students in the two dynamic visualization groups (*drag mode* and *sliders* group) produced more mostly viable summary sheets than the two static groups. Chi-square tests show this difference to be statistically highly significant (p < 0.001). So it can be assumed that students in the two dynamic groups gained more viable knowledge during the intervention than students in the static groups. It might be due to the prepared files students in the dynamic groups had, that less misconceptions or wrong statements are written on their summary sheets. The linked representations available to the dynamic groups that were constantly displayed could have provided even more structure and guidance than the possibility for linked representations that the function plotter group had. Thus the *size of discoveries* (one of the design principles described by MOSSTON 1972) might have been less in the dynamic groups than the static groups even though the task sequence was the same. The fast manipulation in the two

---

[2]In total there were up to five coders, the author, three master-thesis candidates and a student helper, the interrater was always tested against the author. The different coders showed different accordance values, which is not surprising.

dynamic groups could also attribute to a time factor in favour of those two groups and resulted in restructured summary sheets as the students had more time for them.

The overall viability suggests that the *drag mode* and *sliders* groups had gained more viable insights than the static groups. But for a clearer view of the reasons behind these differences the overall viability is not suitable as it is too rough. The reasons for the differences were tried to be elaborated by looking at the content regarding each parameter. Some of these results regarding the three parameters a, b and c, language use and explanations will be presented in the following. Deviating from the structure in the intervention the results will be presented in left to right order of the parameters in the equation $f(x) = a \cdot (x - b)^2 + c$.

11.1.3 Results Regarding Parameter a

If students followed the structure given in the worksheet, parameter a was the second parameter investigated by the students. Contrary to parameter b and c it does not influence the position of the parabola, but rather the slope and therefore the shape of the parabola. For $-1 < a < 1$ the parabola appears wider than the standard parabola through vertical shrinking while for $|a| > 1$ it appears narrower than the standard parabola through vertical stretching.

Some students only identified one of the two, so the inductively developed codes were split for stretching and shrinking. The third category developed for parameter a focussed on the reflection through a negative value of a. Finally it was separately coded if students recognised the special case of $a = 0$, when no parabola but a straight line is graphed. For all four categories it was differed between *not written*[3], *viable* (German original "tragfähig") and *non-viable* (German original "nicht tragfähig").

Students chose to argue the influence with different kinds of statements. They usually did not know the technical terms for a vertical stretch and shrink and thus had to state the influence in their own words. Commonly used expressions were for example the width of the parabola, or the parabola is wider/narrower.

Exemplary statements for all four categories can be found in Table 11.3 (German originals of the examples can be found in Appendix D). The statement "The bigger the value input for a, the wider the parabola is" is a non-viable statement regarding the vertical shrink, as it confuses a shrink and a stretch. So even though the students

---

[3] We cannot assume that the students did not find anything out regarding any category, but rather only know that they did not write it onto their summary sheets.

correctly identify that a influences the shape of the parabola, they describe the influence the wrong way around.

Table 11.3 Examples for the Different Categories Regarding Parameter a. Translated by L. Göbel

Category	Example of a non-viable statement	Example of a viable statement
vertical shrink	The bigger the value input for a, the wider the parabola is.	The width of the opening depends on how much the value differs from zero.
vertical stretch	a. If one changes a, the graph gets wider, the bigger the input value is and vice versa.	If one sets for a a positive value, the parabola gets narrower from small to big values.
Reflection	If one changes (a), the parabola only changes in height not in width.	If a is changed, e.g. a negative value is input, the parabola opens downwards and the vertex point is at the top.
Case $a = 0$	*No non-viable answers present*	If for $a \to 0$ is input, the graph lies on or parallel to the x-axis.

Reasons for the influence of a were only given on a very small number of summary sheets and the results for this categories will be presented in subsection 11.1.7.

All absolute numbers in each category and the percentages for the four categories can be found in Table 11.4. Chi-Square tests show for all four categories the differences to be statistically highly significant ($p < 0.001$). In all four categories the *function plotter* group stated viable answers less often than the other groups.

Also in all groups except the *sliders* group the stretching seemed to be easier than the shrinking with a higher rate of viable answers. The difference between stretching and shrinking is less in the dynamic groups, while in the *function plotter* group 44.8% stated viable answers regarding the stretch and only 20.9% stated viable answers regarding the shrink. The *without visualization* group even outperformed the *sliders* group in the vertical stretch category with a higher percentage of viable statements regarding the vertical stretch. For the *without visualization* group it might be conjectured that students tended to focus on the extreme values of a (so −5 or 5 first), plotted them first and thus realised the vertical stretch as the first influence. The *sliders* group, however, might not move the sliders to the extreme ends of the range and therefore did not identify the vertical stretch as often as the *without visualization* group.

Table 11.4 Results of Parameter a

Category		Without visualiza-tion (N=71)	Function Plotter (N=67)	Drag mode (N=85)	Sliders (N=130)	Total (N=353)
Vertical Shrink	Not written	40 (56.3%)	45 (67.2%)	30 (35.3%)	49 (37.7%)	164 (46.5%)
	Viable	26 (36.6%)	14 (20.9%)	46 (54.1%)	63 (48.5%)	149 (42.2%)
	Non-Viable	5 (7.0%)	8 (11.9%)	9 (10.6%)	18 (13.8%)	40 (11.3%)
Vertical Stretch	Not written	24 (33.8%)	29 (43.3%)	14 (16.5%)	29 (22.3%)	96 (27.2%)
	Viable	45 (63.4%)	30 (44.8%)	60 (70.6%)	77 (59.2%)	212 (60.1%)
	Non-Viable	2 (2.8%)	8 (11.9%)	11 (12.9%)	24 (18.5%)	45 (12.7%)
Reflection	Not written	38 (53.5%)	44 (65.7%)	41 (48.2%)	32 (24.6%)	155 (43.9%)
	Viable	33 (46.5%)	21 (31.3%)	40 (47.1%)	86 (66.2%)	180 (51.0%)
	Non-Viable	0 (0%)	2 (3.0%)	4 (4.7%)	12 (9.2%)	18 (5.1%)
Case $a = 0$	Not written	71 (100%)	67 (100%)	71 (83.5%)	107 (82.3%)	316 (89.5%)
	Viable	0 (0%)	0 (0%)	14 (16.5%)	23 (17.7%)	37 (10.5%)
	Non-Viable	0 (0%)	0 (0%)	0 (0%)	0 (0%)	0 (0,0%)

$p < 0.001$ for all four categories, Cohen's kappa for vertical shrink $\kappa = 0.608$, for vertical stretch $\kappa = 0.485$, for reflection $\kappa = 0.701$, for case $a = 0$ $\kappa = 0.951$.

The low percentage of viable answers in the reflection category by students in *function plotter* group might be due to the students not plotting negative values of a and therefore not seeing reflected parabolas. Likewise, the higher percentage in the *sliders* group in the reflection category might be due to students testing the extreme values of the slider or even just play with it and thus producing a reflected parabola.

If all four categories are taken together, it can be reasoned that the dynamic visualization groups are slightly in favour than the static groups. It is also important that only dynamic visualization groups found the special case $a = 0$ worth noting on the summary sheet. It might be that the static groups did not even identify this special case or only did not find it important enough to write it down. One reason for the difference might be that students in the dynamic groups had to pass the special case dynamically if they were to display a graph with a negative value for a and thus were interested in this anomaly.

In total, the general results regarding parameter a seem to favour the sliders for the reflection and case $a = 0$, but the drag mode for the vertical stretch. The two dynamic groups, especially the *sliders* group might have offered even more implicit guidance through the programming of the file. In the case of the sliders the given range might have hindered the insight into the vertical stretch more than the free manipulation in the drag mode and therefore in the case of the vertical stretch the drag mode is more beneficial. Classical function plotter seem to be the least suitable for investigating the influence of a, one reason might be that students did not plot enough or significant examples to achieve the insight. Thus, the dynamic approaches should be preferred, if a is to be investigated.

11.1.4 Results Regarding Parameter b

In the structure on the worksheet, parameter b was the last parameter to be investigated by the students in the intervention. Both parameters b and c do not influence the shape of the parabola, but only the position of the graph. Parameter b is also the x-coordinate of the vertex point of the graph $f(x) = a(x - b)^2 + c$. If students investigated the formula $f(x) = (x - b)^2$, which was given on the worksheet, b is also the x-intercept. For $b < 0$ the parabola is moved to the left compared to the standard parabola, for $b > 0$ it is moved to the right compared to the standard parabola. The two different directions of movement were coded separately to differentiate if students only identified the influence for positive or negative b. In addition, it was also coded if the students identified the horizontal influence in total. If students identified only movement to the left or right, the horizontal influence was coded as well. Students argued differently but the movement to the left or right was often made explicit as well as describing the position of the vertex points. Examplary statements for the different categories can be found in Table 11.5 (German originals of the examples can be found in Appendix D). In all three categories it was differed between *not written*, *viable* (German original "tragfähig") and *non-viable* (German original "nicht tragfähig"). The statement "$\Rightarrow b$ represents: It moves the vertex ($\rightarrow$ Parabola) $+b$ to the right and $-b$ to the left" is a non-viable statement as it confuses the direction of the movement as well as the sign of the parameter. If the equation is $f(x) = (x + b)^2$ it moves to the left for a positive value for b.

Table 11.5 Examples for the Different Categories Regarding Parameter b. Translated by L. Göbel

Category	Example of a non-viable statement	Example of a viable statement
left & right	$\Rightarrow b$ represents: It moves the vertex ($\rightarrow$ Parabola) $+b$ to the right and $-b$ to the left.	b determines the position on the x-axis. (with minus to the left, with plus to the right).
horizontal	$\Rightarrow b$ represents: It moves the vertex ($\rightarrow$ Parabola) $+b$ to the right and $-b$ to the left.	If one has the formula $f(x) = (x - b)^2$ and changes b, the position on the x-axis changes

Table 11.6 Results of Parameter b

Category		Without visualization (N=71)	Function Plotter (N=67)	Drag mode (N=85)	Sliders (N=130)	Total (N=353)
right	Not written	54 (76.1%)	46 (68.7%)	47 (55.3%)	67 (51.5%)	214 (60.6%)
	Viable	13 (18.3%)	10 (14.9%)	22 (25.9%)	51 (39.2%)	96 (27.2%)
	Non-Viable	4 (5.6%)	11 (16.4%)	16 (18.8%)	12 (9.2%)	43 (12.2%)
left	Not written	54 (76.1%)	50 (74.6%)	49 (57.6%)	68 (52.3%)	221 (62.6%)
	Viable	13 (18.3%)	6 (9.0%)	20 (23.5%)	50 (38.5%)	89 (25.2%)
	Non-Viable	4 (5.6%)	11 (16.4%)	16 (18.8%)	12 (9.2%)	43 (12.2%)
horizontal	Not written	22 (31.0%)	6 (9.0%)	1 (1.2%)	8 (6.2%)	37 (10.2%)
	Viable	43 (60.6%)	47 (70.1%)	68 (80.0%)	114 (87.7%)	272 (77.1%)
	Non-Viable	6 (8.5%)	14 (20.9%)	16 (18.8%)	8 (6.2%)	44 (12.5%)

$p < 0.001$ for all three categories, Cohen's kappa for left $\kappa = 0.676$, for right $\kappa = 0.555$, for horizontal $\kappa = 0.682$ after majority decision.

Table 11.6 shows the exact percentages for all three categories split into the intervention groups. It is evident that the two static groups (*without visualization* and *function plotter*) split their results less often into transformation to the left and right than the dynamic groups as a lower number of left and right transformations were coded than in the two dynamic groups, while this difference is not as big in the horizontal category. The two static groups mostly only identify the horizontal influence. In the *without visualization* group there is a rather low (less then 9%) of non-viable statements regarding the horizontal transformation, but nearly a third of students did not write anything about a horizontal transformation. So if students identify the influence they do so mostly viably. The rather high number of no statements might be due to students not attempting to sketch these graphs, maybe due to time constraints. In the *drag mode* group a relatively high percentage of nearly a fifth of students identified the right, left and horizontal influence in a non-viable way, this was often the case, when students stated that the parabola moved to the left if parameter b was positive and vice versa. This problem has already been described in the literature, for example in ZAZKIS et al. (2003). This confusion might stem from the displayed function equation where, if parameter b is negative, e.g. -1 the equation is simplified to $f(x) = (x + 1)^2$. Surprisingly, the *function plotter* group show a rather high percentage of a fifth of all students with non-viable statements regarding the horizontal transformation. Including in this number are statements, where the direction of the horizontal transformation is stated the wrong way around. As students in these groups explicitly had to input the values into the function equation, the confusion described by ZAZKIS et al. (2003) (that e.g. the graph of $f(x) = (x + 1)^2$ is moved one unit to the left compared to the standard parabola, even though the plus sign might be associated with movement to the right) should have been prevented.

87.7% of students in the *sliders* group identified the horizontal influence in a viable way, thus for parameter b using sliders seems to be the most appropriate way to achieve a viable knowledge of the influence. This preference might be explained through the design of the software in the sense that the value of parameter b is displayed above the sliders and not input and then simplified into the function equation and therefore avoiding the confusion possibly existing in the *drag mode* group. The differences between the groups are statistically highly significant as shown by a Chi-Square test ($p < 0.001$). In general identifying the influence of parameter b as movement to the left and right seems to be the most difficult for students, while only identifying the horizontal influence is easier.

11.1.5 Results Regarding Parameter c

Parameter c was thought to be the easiest of the three parameters, as it only influenced the vertical position of the parabola. Therefore it was decided that in the equation $f(x) = a \cdot (x - b)^2 + c$, c was the first parameter intended to be investigated by the students. It is also the y-coordinate of the vertex point. In the pre-structure given on the worksheet, students were asked to investigate an equation of the form

$f(x) = x^2 + c$, where then c also gives the y-intercept. For $c > 0$ the parabola is above the standard parabola, while for $c < 0$ it is below the standard parabola. The transformed parabola can be obtained by adding the constant value of c to each y-value of the standard parabola.

As with parameter b the two different directions of movement were coded separately, and it was also coded if students wrote statements about the vertical transformation in general. If students stated something about only one direction of movement, the vertical transformation category was also coded. In all three categories it was, as with parameter b, differed between not written, viable and non-viable answers. The statement "The graph moves depending on what is input, with negative numbers upwards and with positive downwards" is a non-viable answer regarding both the upwards and downwards category. Even though the students identify a vertical influence, they confuse the direction of the movement and state it the wrong way around. Examples for the other categories and codings can be found in Table 11.7 (German originals of the examples can be found in Appendix D).

Table 11.7 Examples for the Different Categories Regarding Parameter c. Translated by L. Göbel

Category	Example of a non-viable statement	Example of a viable statement
upwards & downwards	1. The graph moves depending on what is input, with negative numbers upwards and with positive downwards.	c changes the height of the vertex point on the y-axis, c positive: vertex point moves to the positive, c negative: Vertex point moves to the negative
vertical	For the function $f(x) = x^2 - c$, the vertex point is the lowest with negative numbers, because one calculates $x^2 - c$.	$c = c$ is the y-value of the vertex point.

As predicted, parameter c seems to be the easiest to comprehend by the students as shown by the relatively higher percentages of viable answers compared to parameter b (for all percentages see Table 11.8). This supports the literature, where it was also described that vertical transformation is easier than horizontal (see for example EISENBERG & DREYFUS 1994 and BAKER et al. 2000). This difference could also be due to the order in which the parameters were investigated as it is the parameter all students investigated first, if they followed the structure. In all four groups the percentage of non-viable answers is rather low, this supports the theory that it is the easiest influence to identify correctly. Also in all groups, more than 75% identified

Table 11.8 Results of Parameter c

Category	p		Without visual- ization (N=71)	Function Plotter (N=67)	Drag mode (N=85)	Sliders (N=130)	Total (N=353)
upwards	$p =$ 0.004	Not written	51 (71.8%)	46 (68.7%)	40 (47.1%)	72 (55.4%)	209 (59.2%)
		Viable	18 (25.4%)	19 (28.4%)	42 (49.4%)	52 (40.0%)	131 (37.1%)
		Non- Viable	2 (2.8%)	2 (3.0%)	3 (3.5%)	6 (4.6%)	13 (3.7%)
downwards	$p =$ 0.017	Not written	49 (69.0%)	54 (80.6%)	49 (57.6%)	74 (56.9%)	226 (64.0%)
		Viable	20 (28.2%)	13 (19.4%)	31 (36.5%)	50 (38.5%)	114 (32.3%)
		Non- Viable	2 (2.8%)	0 (0%)	5 (5.9%)	6 (4.6%)	13 (3.7%)
vertical	p=0.005 by Fisher's exact test	Not written	10 (14.1%)	8 (11.9%)	1 (1.2%)	4 (3.1%)	23 (6.5%)
		Viable	55 (77.5%)	53 (79.1%)	79 (92.9%)	112 (86.2%)	299 (84.7%)
		Non- Viable	6 (8.5%)	6 (9.0%)	5 (5.9%)	14 (10.8%)	31 (8.8%)

Cohen's kappa for upwards $\kappa = 0.603$, for downwards $\kappa = 0.604$, for vertical $\kappa = 0.828$ after majority decision.

the vertical influence viably. This is a very good result for the shortness of the intervention overall, but the results of the *drag mode* group are even better, where 92.9% of students identified the vertical influence in a viable way. Also only one student in the *drag mode* group did not write anything regarding the influence of c on the summary sheet. Again students less often split their insights into upwards and downwards movement, which leads to lower percentages of viable answers.

In total, the two dynamic groups seem even more appropriate approaches for investigating the influence of parameter c, while the *without visualization* group seems to be the least appropriate version. It can be conjectured that the programming of the files for the two dynamic groups foster conceptualization. In the *drag mode* group the function equation directly changes as soon as the students move the parabola, while the *sliders* group intentionally moves the slider for c and see the resulting change in the graph. The less appropriateness of the *without visualization* group might be due to students not sketching enough or significant examples to visualize the influence correctly. However, to test for statistically significance the non-viable answers had to be left out for the upwards and downwards category, as otherwise the expected count in four cells was below 5 and therefore the requirements for the chi-square test were not given (RASCH et al. 2014, p. 129). After this adjustments, the differences between the intervention groups were statistically significant. The differences in the vertical category are statistically significant (p = 0.005). The results regarding the different parameters show an advantage of the dynamic groups over the static groups. However, not only the viability is interesting, but also the different uses of language by the students.

11.1.6 Results Regarding Language Use

The summary sheets are products of students' writing and the analysis of these therefore focusses on the language use of students. How students use everyday language and technical terms to describe mathematical concepts can give hints towards the depth of their understanding. It was therefore decided to code the different technical terms used by the students as well as the viability of the context used. A total of thirty different technical terms were coded separately while also a category for other technical terms that were not included in the thirty explicitly coded was included. Interesting results showed for the technical terms *y-axis*, *x-axis* and *slope*. In all three categories it was again differed between *not used*, *viable* (German original "tragfähig") and *non-viable* (German original "nicht tragfähig"). Surprisingly, for the *x*-axis and *y*-axis category, there were no statements on the summary sheets, where the terms were used in a non-viable way. For all other examples see Table 11.9 (German originals examples see Appendix D). The non-viable example for the *slope* is non-viable as the students try the rise-over-run method on the parabola, which probably stems from over-generalizing from the linear functions (see also ZASLAVSKY 1997, p. 32 f. or ELLIS & GRINSTEAD 2008).

Contrary to the expectations, the terms *x*- and *y*−axis were used more in the experimental groups with the highest percentage in the *drag mode* group (see Table 11.10). Chi-square tests show that the differences are statistically significant (p < 0.001) in both categories. Reasons for these differences might originate in the different technology used. Students in the *without visualizations* group sketched the graph per hand and therefore might not have deemed it necessary to describe the movement, while the other three groups had the technological visualizations and thus needed to interpret the resulting graphs verbally. This might have led to the

Table 11.9 Examples for the Different Categories Regarding Technical Terms. Translated by L. Göbel

Category	Example of a non-viable statement	Example of a viable statement
x-axis	*No non-viable answers present*	b changes the height of the vertex point on the x-axis.
y-axis	*No non-viable answers present*	c changes the height of the vertex point on the y-axis.
slope		a influences the slope

a is the slope

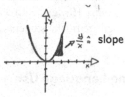

slope

Through the slope
varies the width the
parabola. With a
negative slope the
opening of the parabola
turns downwards.

necessity of using more technical terms to describe the changes. Students in the *drag mode* group also moved the parabola along the axes to the new positions and this might have been prompted to use these terms to describe the changes.

The term *slope* was not widely used in all four groups, with three quarters of all students not writing the term slope on their summary sheet. If students used the technical term *slope* they often used it in a non-viable way. This happens for example, when students over-generalize their knowledge regarding the slope of linear functions and state that a in $f(x) = a \cdot (x-b)^2 + c$ is the slope. This problem has been described in the literature as a main obstacle regarding quadratic functions (see section 2.2.3 and ZASLAVSKY 1997, BILLS 2001, ELLIS & GRINSTEAD 2008). An example how the stereotyping of parameters influences the process is described in subsection 12.3.3. The non-viable use occurs more often in the two static groups and the highest percentage of non-viable use is in the *without visualization* group. The three experimental groups might overcome the illusion that a is the constant slope due to the higher number of examples they could have considered. Also the

Table 11.10 Results Technical Terms

Category		Without visualiza- tion (N=71)	Function Plotter (N=67)	Drag mode (N=85)	Sliders (N=130)	Total (N=353)
x-axis	Not written	55 (77.5%)	27 (40.3%)	28 (32.9%)	53 (40.8%)	163 (46.2%)
	Viable	16 (22.5%)	40 (59.7%)	57 (67.1%)	77 (59.2%)	190 (53.8%)
	Non- Viable	0 (0,0%)	0 (0%)	0 (0%)	0 (0%)	0 (0%)
y-axis	Not written	45 (63.4%)	27 (40.3%)	25 (29.4%)	58 (44.6%)	155 (43.9%)
	Viable	26 (36.6%)	40 (59.7%)	60 (70.6%)	72 (55.4%)	198 (56.1%)
	Non- Viable	0 (0%)	0 (0%)	0 (0%)	0 (0%)	0 (0%)
slope	Not written	43 (60.6%)	44 (65.7%)	75 (88.2%)	106 (81.5%)	268 (75.9%)
	Viable	0 (0%)	4 (6.0%)	6 (7.1%)	10 (7.7%)	20 (5.7%)
	Non- Viable	28 (39.4%)	19 (28.4%)	4 (4.7%)	14 (10.8%)	65 (18.4%)

$p < 0.001$ for all three categories, but only when viable answers were left out in the case of slope, Cohen's kappa for x-axis $\kappa = 0.854$, for y-axis $\kappa = 0.913$, for slope $\kappa = 0.692$

without visualization group might have needed the concept of slope to sketch the graphs while the two dynamic groups did not view the slope but rather the complete parabola and thus focussing more on the Grundvorstellung object than the control group. The problem of stereotyping the parameter and to generalize from linear functions has been described in the literature as well (ZASLAVSKY 1997, BILLS 2001).

The results presented above regarding all three parameters only code if the influence is identified and stated on their summary sheets. However, for conceptualization it is also important to be able to explain the parameter influence. The results regarding the explanations will be presented in the next subsection.

11.1.7   Explaining—a Special Challenge

The aim was to initiate deeper reflection on the investigation by implementing the third part of the intervention task, where students were intended to find explanations for the transformations discovered in part two of the intervention (see also Figure 7.17 and Figure 7.18). It was anticipated that finding explanations would not be easy and therefore the students would profit from some scaffolding. For this, fictitious statements were implemented for parameter b and c, as well as a function table for two values of c. For parameter c the fictitious statements highlighted a typical misconception that the graph gets narrower (e.g as described by PINKERNELL 2015). In order to ascertain if explaining was present, for all categories described in subsections 11.1.3, 11.1.4, 11.1.5 it was also coded if they explained it. Again it was differed between *not written*, *viable* and *non-viable*. The results were not as expected and showed that only very few students wrote any explanations down at all. Due to the low number of overall explanations it was not sensible to differentiate between the intervention groups. Of the very few explanations on the summary sheets many of them were then non-viable as well, so it can be concluded that the intervention did not foster explanations as good as was intended. Exact percentages can be found in Table 11.11. The absence of the explanations might be explained by the shortness of the intervention. Also, students might not have succeeded to get

Table 11.11 Frequency Explanations of the Different Transformations Described Above

Parameter	Category	N	Viable	Non-viable	Not written
a	vertical shrink	353	7 (2.0%)	12 (3.4%)	334 (94.6%)
	vertical stretch	353	8 (2.3%)	12 (3.4%)	333 (94.3%)
	special case $a = 0$	353	6 (1.7%)	8 (2.3%)	339 (96.0%)
	reflection	353	0 (0%)	0 (0%)	353 (100%)
b	left	353	0 (0%)	2 (0.6%)	351 (99.4%)
	right	353	4 (1.1%)	6 (1.7%)	343 (97.2%)
	horizontal	353	2 (0.6%)	14 (4.0 %)	337 (95.5%)
c	upwards	353	12 (3.4%)	3 (0.8%)	338 (95.8%)
	downwards	353	4 (1.1%)	2 (0.6%)	347 (98.3%)
	vertical	353	8 (2.3%)	10 (2.8%)	335 (94.9%)

far enough with their conceptualization of parameters in order to be able to explain the influence or they did not understand the wording of a task as a request for an explanation.

11.1.8 Conclusion General Results Summary Sheets

The overall structure category shows that two-thirds of the students were able to restructure the gained information into summary sheets. This re-grouping needs a certain degree of conceptualization in order to structure it, so it can be reasoned that the intervention leads to new knowledge including conceptual elements.

This can be supported by the high percentages over all groups stating the influence of parameter c in a viable manner, thus the intervention succeeds in developing new knowledge. The dynamic groups write more viable statements regarding parameter a and b, with advantages for the *sliders* group regarding b and *drag mode* group for parameter a and c. In general the two dynamic groups seem to be better suited than the two static groups. If the students are only split into these two groups, all categories described in subsections 11.1.3, 11.1.4, and 11.1.5 show significant differences ($p < 0.005$) between the dynamic and static groups (exact tables see Appendix E). So even though there are differences between the *drag mode* and *sliders* group, if taken together it can be reasoned that they seem more suitable for investigation of the parameter influence than the two static groups. But the summary sheets only show what the students wrote down and the categories described above do not analyse each statement in detail, also the coding only took the viability of answers into consideration.

11.2   Students' Reasoning Concerning Parameter c

In order to identify the different ways students argue their insights, answers concerning parameter c were analysed in a more detailed manner. The nature of the technology used appeared to have an influence on the wording used to describe the influence. The development of the coding manual for this detailed analysis is described above in subsection 8.3.2. Not all summary sheets contained statements regarding parameter c, while others contained more than one statement. Statements were splitted if necessary in order to identify most ways of arguing by the students. As described in subsection 8.3.2 answers could be clustered into the five categories *y-intercept, y-axis, parabola shape, vertex point, slope*[4]. It was also differed between *dynamic* and *static* statements, for example the statement "*c* changes the y-intercept" can be seen as a dynamic statement as it focusses on the change, while the statement "*c* is the y-intercept" is more static. It was conjectured that the dynamic groups (*drag mode* and *sliders* group) would use more dynamic statements as their technology showed the actual movements of the parabola, while the static groups (*without visualization* and *function plotter*), would use more static statements. The two static groups only saw a graph being substituted by another and not moved dynamically.

While dynamic and static statements were expected, it was not deliberated or conjectured how students would formulate their insights. But the ways the students formulated their results were not surprising. Arguing with the *y-intercept* stems most likely from the structure given by the worksheet as it was suggested that students use the formula $f(x) = x^2 + c$. In this formula c is actually the y-intercept. For the *without visualization* group the y-intercept is also often computed to sketch a graph and therefore might be then used to describe their influence.

Using the y-axis to explain the influence also is not surprising as describing the direction of the movement is often then referred to the axes. Not as obvious are the statements regarding the parabola shape, however, a closer look at the worksheet might give a reason. In order to avoid the illusion that the graph changes width when moved upwards described by PINKERNELL (2015, p. 2532), the two statements regarding the parabola shape were implemented to make the students aware of this illusion. It then follows that students might use this to argue the influence.

Using the slope in order to explain the influence often points to a over-generalizing of linear functions and the attempt by the students to match each parameter in the quadratic functions to one in linear functions (see for example Charlotte in subsection 12.3.3). Finally, using the vertex point for describing the influence of parameter c, which is always the y-coordinate of the vertex point in the context of this intervention seems to be an obvious choice.

Some students using the vertex point for their argument also then described the movement along the y-axis, these answers were then double coded into both categories. For the statistical evaluation of the coding presented in this section, it was decided to not count the number of students who participated in designing the summary sheets, but the number of statements regarding parameter c that were included in the analysis. This decision was due to some students stating more than one separate statement on their summary sheets, which were then coded separately, while others did not write any statement regarding c. The double coding described

---

[4]Sometimes the category was only implicitly, so for example a movement up and downwards was coded as y-axis.

above the number of statements in the content related categories *y-intercept, y-axis, parabola shape, vertex point, slope* exceeds the number of statements in the dynamic/static coding.

Table 11.12 Dynamic and Static Statements Regarding Parameter c

Group	Number of statements	dynamic	static
Without Visualization	37	18 (48.7%)	19 (51.3%)
Function plotter	38	14 (36.8%)	24 (63.2%)
Drag mode	45	27 (60.0%)	18 (40.0%)
Sliders	68	41 (60.3%)	27 (39.7%)
Total	188	100 (53.2%)	88 (46.8%)

Table 11.12 shows the results regarding the dynamic and static statements split into the different intervention groups. The results support the hypothesis that the dynamic groups formulate more dynamic answers as around 60% of statements in the *drag mode* and *sliders* group are of a dynamic nature. Surprisingly, the *without visualization* groups' statements are of a dynamic nature in 48.7% of the statements. The lowest number of dynamic statements are visible in the function plotter group with only 36.8% of statements. It could be that the *function plotter* group only operates on the parameter as a placeholder (as described by DRIJVERS 2003, see subsection 2.3.2) in the static entering of syntax. Thus the process of plotting one function after another focusses more on the static objects rather than on a dynamic change of function. In the *without visualization* group however, as the students actively sketch the function per hand some degree of dynamic view might be retain.

The way of modification of the graphs in the two dynamic groups corresponds more to the parameter as a changing quantity (see DRIJVERS 2003, subsection 2.3.2), which encorporates a more dynamic view and might therefore result in more dynamic statements.

The results for the content coding of the statements are not as clear as the results concerning the dynamic and static statements (see Table 11.13 for exact results). But preferences in the reasoning emerge there as well. Reasoning with the slope only occurs in a very small number of statements, while vertex point and y-axis seem more popular to be used in the reasoning. The *y*-intercept is used in 34% of all statements in the *without visualization* group, while in the *sliders* group it is only present in 14% of statements. Supporting the results regarding the *y*-axis presented

Table 11.13 Content Codings of Statements Regarding Parameter c

Category	Without visualization (N=38)	Function Plotter (N=42)	Drag mode (N=51)	Sliders (N=78)	Total (N=209)
slope	2 (5%)	3 (7%)	1(2%)	0 (0%)	6 (3%)
vertex point	14 (37%)	12 (29%)	12 (24%)	28 (36%)	66 (32%)
parabola shape	3 (8%)	6 (14%)	7 (14%)	8 (10%)	24 (11%)
y-axis	6 (16%)	11 (26%)	23 (45%)	31 (40%)	71 (34%)
y-intercept	13 (34%)	10 (24%)	8 (16%)	11 (14%)	42 (20%)

in subsection 11.1.6, the two dynamic groups used the y-axis to reason in 45% and 40% of statements in the *drag mode* and *sliders* group respectively.

Conclusion Students' Reasoning for Parameter c

The analysis of statements regarding parameter c shows that students in the *drag mode* and *sliders* group tend to use more dynamic language than students in the static groups. A reason could be a different focus of Grundvorstellungen, while working with the functions. It could be argued, that the *without visualization* group focusses mostly on the Grundvorstellung mapping as they have to compute values for each function and then sketch the graphs from these values. The *function plotter* group has the static mapping of one function equation to one graph and thus might focus on certain points again using the Grundvorstellung mapping, while the two dynamic groups focus on the dynamic changes of the graph. The most likely reason for the differences stems from the way the different technologies are designed. If the change is dynamic, it is understandable that students use dynamic language, while students who only see static change use static language.

The reasoning used by the students is often based on the vertex point and the y-axis, the later especially by students of the two dynamic groups, while a third of statements in the *without visualization* group are based on the y-intercept and and a third based on the vertex point. This again might stem from the design of the different intervention groups. For the *without visualization* group the y-intercept and the vertex point are most likely important points for the process of sketching graphs and therefore the Grundvorstellung mapping is most relevant. The *drag mode* group drags the function at the vertex point and then moves it along the y-axis, thus the focus is more on the movement along the y-axis in the *drag mode* group and this might explain the relatively high number of statements (45%). The same reasoning can be applied to the *sliders* group, where students observe the change of the vertex point along the y-axis if they move the slider for c, while keeping the value for b at zero. This might lead to a focus on the Grundvorstellung object.

11.3   Conclusion Analysis of Summary Sheets

The summary sheets show the diverse ways in which students generated and formulated their results in the intervention. It can be concluded that the intervention does lead to conceptualization of parameters in all groups, so the Guided Discovery seems to be a good way to achieve this goal, regardless in which group students were. The implemented guidance on the worksheet triggers the process of discovery by the students and through this the Guided Discovery leads to summary sheets that include many key elements and important points regarding the conceptualization of parameters.

Beside this general success of the intervention there are several differences between the different groups that were identified. Thus, it can be induced that the technology also facilitates the conceptualization in different ways. The dynamic groups (so *drag mode* or *sliders* group) outperform the static groups (so *without visualization* or *function plotter* group) regarding all parameters visible through a higher number of viable statements. These differences are statistically significant (see Appendix E for tables of results split between dynamic and static). The dynamic groups were given a file for the used technology that included linked multiple representations that could not be turned off. Linked multiple representations have been shown to be beneficial for learning (see section 3.2) and they might have decreased the *size of discoveries* (see section 4.2) necessary to occur in the investigation. In one of the static groups, namely the *function plotter* group the linked function tables were not visible all the time, though they could be displayed through the software, but students might not have used this feature.

But the division only into static and dynamic is too rough, as there are differences between the different intervention groups. However, not only are differences between the four intervention groups visible, these differences are also dependent on the parameter investigated. The drag mode seems to be better suited than the other three approaches for investigating parameter a and c, while the use of sliders evoke the most viable statements regarding the horizontal transformation through parameter b. It becomes evident, however, that even in the parameter influences itself there are differences between the four intervention groups. So for parameter a, if the vertical stretch is to be investigated the drag mode seems the most beneficial, while for the reflection at the x-axis through the parameter a sliders are deemed to

Table 11.14 Overview Over the Results Split Between the Parameters

Parameter	Category	*Without visualization*	*Function Plotter*	*Drag mode*	*Sliders*
a	Vertical Shrink				
	Vertical Stretch				
	Reflection				
	Case $a = 0$				
b	Right				
	Left				
	Horizontal				
c	Upwards				
	Downwards				
	Vertical				

For the color of the cell the detailed results described above are taken into account. Green represents the most suitable way to investigate the influence, yellow is undecided, red is the least suitable way as described in section 11.1.

be the most appropriate approach. Table 11.14 visualizes the results of the summary sheet analysis in regard to which approach is the most suitable for which parameter influence.

The question arises, why the technological approaches differ in the suitability for the different parameter influences. The drag mode offers "direct and sensual manipulations of graphs" (SEVER & YERUSHALMY 2007, p. 1517). This might be helpful for the discovery of the vertical transformation through parameter c and the vertical stretch and shrink through parameter a. For parameter c moving a slider left and right and observing a vertical movement might have presented not such a clear view or even resulted in a overburdening cognitive load (through the greater number of linked representations, see BOERS & JONES 1994 and section 3.2) and therefore obscured the insights more than the direct movement in the drag mode.

In comparison to the transformation through parameter c, transformation through change of parameter b has been described as more difficult due to the direction of the change and the non-intuitive direction of transformation (e.g. by EISENBERG & DREYFUS 1994). It is also evident to some degree in this study as the influence of parameter c has been identified in a viable way more often than the one through b. The connection of moving a slider right and left and at the same time observing the graph moving horizontally, might be more beneficial for the learning about parameter b than dragging at the vertex point of the parabola. The slider might have functioned as a process constraint and reduced the amount of choices needed (as described by DE JONG & LAZONDER 2014). The slider could also act as a direct external representation of the parameter as a changing quantity (for example described by DRIJVERS 2003) and thus foster the insight into the influence.

Surprisingly, students in the *without visualization* group wrote more viable statements on their summary sheets regarding vertical shrink, vertical stretch and reflection of the parabola than students in the *function plotter* group. It might be reasoned that students in the *without visualization* groups chose more significant examples, for example choosing to sketch the extreme values of -5 and 5 first and therefore gaining more insight, than students in the *function plotter* group. Students in the *function plotter* group had to actively choose what graph to plot and might not have chosen significant enough examples to observe the influence or decided to plot many functions at the same time and then could not differentiate between the examples.

In the case of explanations there are no differences between the different intervention groups nor the parameters, as in all cases there are very little explanations present (see subsection 11.1.7). This does not mean that students did not succeed at all in finding explanations as the video analysis will show in the next chapter (see subsection 12.2.4), but the video analysis also shows that for some students the task regarding finding explanations did not evoke any need to explain the influences (see subsection 12.3.4). So the reasons for the small number of explanations cannot be cleared up completely.

It became evident in the summary sheets analysis that students in the *drag mode* and *sliders* groups use more dynamic language in the description of the influence through c and it can be conjectured to be the same for the other two parameters b and c. This was, however, not investigated in the frame of this thesis. Regarding the statements describing the influence of parameter c it has been shown that students' statements fall into one of five content categories with a preference to argue using the vertex point and y-axis. The design of the worksheet might have emphasized the position of the graph in the coordinate system compared to the origin and thus the tendency to argue using y-axis for the influence of c would be not surprising.

The summary sheets mainly show what students wrote in part four of the intervention, so it could be objected that students in the control group and function plotter groups write lesser results onto their summary sheets due to time constraints and not because they found out less.

This cannot be cleared up by an analysis of summary sheets as they do not show the processes undertook by the students in the different groups. For this the videos of the intervention were analysed and the results are shown in the following Chapter 12.

Videos

12

The results of the summary sheet analysis show differences between the three experimental groups and the control group. They can point to the conclusion that the dynamic visualizations used in the *sliders* and *drag mode* groups are more suitable for the use in the conceptualization of parameters of quadratic functions than the static visualizations used in the *without visualization* and *function plotter* groups. These results, however, do not show how the technology can facilitate this. More insight into this question can be gained through analysis of the videos. Thus, starting with two examples of how the Guided Discovery progressed, it will be shown how the technology can foster and change the process of conceptualization. Then, illustrating different affordances and constraints of the technology and Guided Discovery using case studies will be presented in order to deepen the insight into how the conceptualization of parameters takes place.[1]

12.1 Two Examples of a Guided Discovery Process

The Guided Discovery processes of the different students proceed in various ways depending on the students' choices during the Guided Discovery, the used technology as well as their general mathematics skills. Insight into processes of two groups will be presented in the following in order to retrace some of these various ways using graphic timelines in which the processes are visualized. Through this visualizing of the actions the students take, common sequences and interdependen-

---

[1]Transcripts of the processes presented in this chapter were translated by the author. It was taken care to keep to the students' language as closely as possible as well as trying to not lose any meaning through phrasing.

© The Author(s), under exclusive license to Springer Fachmedien Wiesbaden GmbH, part of Springer Nature 2021
L. Göbel, *Technology-Assisted Guided Discovery to Support Learning*,
Essener Beiträge zur Mathematikdidaktik,
https://doi.org/10.1007/978-3-658-32637-1_12

cies become visible, so for example the sequence of using technology and stating a hypothesis afterwards can be identified a number of times in the two examples.

The two examples were chosen to show, how students work with a static and a dynamic visualization in the Guided Discovery and were videographed in a *function plotter* and a *sliders* group respectively. Many of the common potentials and constraints of the approach described in sections 12.2 and 12.3 are visible in one or both of the two example processes. While the students in the *function plotter* group video, Charlotte and Noah[2] can be considered middle achievers with pretest scores of 30 and 33 out of 56 points respectively, the students in the *sliders* group video, Tom and Iwan, were high achievers with pretest scores of 52 and 45.5 points respectively.

12.1.1 An Example Using Function Plotters

Charlotte and Noah were in the *function plotter* group and thus were given an iPad with a blank page on the TI-Nspire CX CAS App. This process is chosen to illustrate how the Guided Discovery using the static visualization through function plotters progresses. The two students worked on their own on the tasks for a total of 58 minutes and 44 seconds, during which they were very concentrated on the task and did not stray off topic. While working, they systematically followed the structure of the worksheet and, therefore, investigated the different parameters one after the other.

To visualize the way in which the process progresses, the timeline in Figure 12.2 was created in order to identify common sequences and interdependecies between the technology and insights. It includes the graphed equation and important points of the process. The height of the phase rectangular is proportional to the time (the number in the phase corresponds to the descriptions in section F.1). The color of the phase stands for the parameter investigated in this phase, while the symbols left of the phase rectangulars visualize what actions the students take (from left to right). The identified actions can be clustered into technological aspects for example technical problems, aspects regarding the working organisation, e.g. formulating the summary sheet, aspects regarding content related so insights, explanations etc. and interaction with the researcher.

For example in Phase 8 in Figure 12.2 the students first input the function $f(x) = x^2 + (-2)$ and state a hypothesis, regarding the upwards movement. They then input a new function, however, a technical problem occurs due to forgetting a plus sign.

---

[2] All names in this chapter were altered.

They correct this mistake and plot the function $f(x) = x^2 + (-1)$ and thus achieve some insight. This phase of the Guided Discovery is represented through the symbol sequence in Figure 12.1:

Figure 12.1 Example of Symbol Sequence Visualizing Guided Discovery Process

The visualization through the timeline and analysis of the occurring processes enables to gain more insight into the learning pathways than the summary sheet analysis described in Chapter 11 provided. At first glance it can be deduced by the number of symbols in the timeline that especially in the first half of the intervention the students' process is densely packed with many different actions undertaken by the students (Figure 12.2).

The students gather valuable and many viable insights through their investigation of various examples, both examples that were given on the worksheet as well as examples of their own choosing (e.g. $f(x) = -10x^2$ in Phase 30). The length of their investigation differs for the parameters, Charlotte and Noah spend the longest time on Parameter a. They identify the influence of c as vertical movement, they discuss the special case of $a = 0$ and the influence through a including that negative values of a cause a reflection at the x-axis. They also recognize that b is the position of the parabola on the x-axis. Explanations for the influences, however, are not present. In the process of gaining viable insights, some misconceptions or non-viable insights are also present.

But before the students are able to use the technology to gain their insights, they encountered some technological obstacles at the beginning (see Phase 6 and 7). To overcome the technological obstacles, they asked for help a number of times, but work mostly independently.

The use of technology eventually fosters the discoveries as the students use the technology to plot the examples and compare them.

A common use of the technology (see for example Phase 8 or 10) in their process is the use of the technology followed by the statement of a hypothesis and again a use of technology with a resulting statement of an insight. They also use the technology to overcome the misconception that the graph is narrower, if parameter c is changed (in Phase 25). This misconception stems from the wrong interpretation of the graphical information (as described e.g. by GOLDENBERG 1988, PINKERNELL 2015 and subsection 2.2.3). For this, Noah uses the technology to explain to Charlotte, why their earlier statement that the graph is narrower is not correct (see subsection 12.2.3

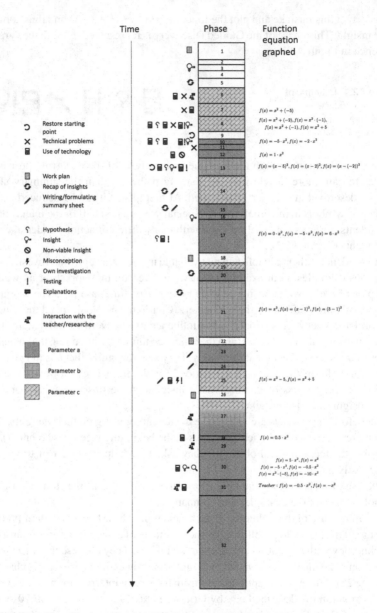

Figure 12.2 Timeline for a Guided Discovery Process Using Function Plotters

for an analysis of this communication including technology). They then overcome the misconception that c influences the shape of the parabola and correctly state the influence of c.

During their investigation Charlotte shows strong stereotyping for the parameters based on her knowledge of linear functions. This first occurs, when she identifies parameter a as m, so the slope in linear functions. It occurs again, while formulating their results for their notes, where she compares the parameters given in the intervention with the ones of linear functions (see subsection 12.3.3 for an analysis of this sequence).

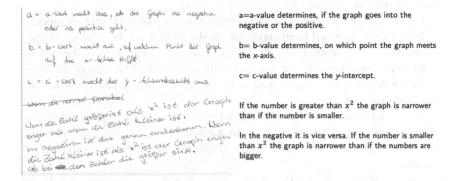

a=a-value determines, if the graph goes into the negative or the positive.

b= b-value determines, on which point the graph meets the x-axis.

c= c-value determines the y-intercept.

If the number is greater than x^2 the graph is narrower than if the number is smaller.

In the negative it is vice versa. If the number is smaller than x^2 the graph is narrower than if the numbers are bigger.

Figure 12.3 Summary Sheet Produced by Charlotte and Noah. Translated by L. Göbel

When one looks at the final summary sheet they produced, it becomes clear that insights that occurred in the investigation are not written on the summary sheets. For Charlotte and Noah this is for example the insight that the graph is a straight line for $a = 0$, they specifically investigate this special case on Charlotte's request (in Phase 17), but do not write it on their final summary sheet they produce in the second half of their investigation (Phase 22 and following, see Figure 12.3).

The insight view into the learning pathway by Charlotte and Noah shows that the summary sheet analysis in Chapter 11 does not provide a complete view and the video analysis provides further insight into how the use of technology progresses and what misconceptions occur. A second example of a complete progress using a dynamic visualizations namely sliders will be presented in the next section.

12.1.2 An Example Using Sliders

The second example illustrates the ways the Guided Discovery process can take, while using sliders. Tom and Iwan achieved the highest and third highest pretest score in their class. They were given an iPad with the prepared file with sliders on the TI-Nspire CX CAS app. The two boys work for 50 minutes and 6 seconds on the tasks without straying from the topic too far. As with Charlotte and Noah, this is a good result for two Year 9 students working completely on their own. So the Guided Discovery enables productive working on their own. This is also evident in the videos not presented in this section as well as being observed with the students that were not videographed during the intervention.

Figure 12.5 shows a timeline, which visualizes their Guided Discovery process, with the values Tom and Iwan chose for the sliders in the right column. As with the timeline in subsection 12.1.1, the height is proportional to the time (the number in the phase corresponds to the descriptions in section F.2). The color of the phase stands for the parameter investigated in this phase, while the symbols represent the action that takes place.

As with Charlotte and Noah, if one looks only at the summary sheet produced by Iwan and Tom (see Figure 12.4), there are a number of aspects that are not visible. The analysis of the video provides more insight into the learning pathways of the students.

The summary sheet (see Figure 12.4) shows that the two students identify the influence of all three parameters viably, but omits for example that the two boys even

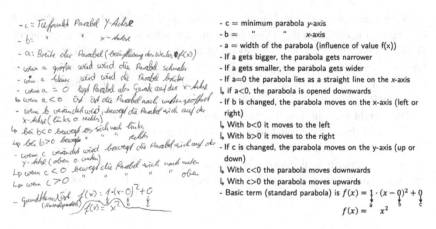

Figure 12.4 Summary Sheet Produced by Tom and Iwan. Translated by L. Göbel

attempt explanations. However, the summary sheet does show the amount of details the two boys succeeded in identifying. For all three parameters the influences are described and for the parameters b and c differed into positive and negative values. The special case $a = 0$ is also described.

Contrary to Charlotte and Noah, Tom and Iwan spent the most time on investigating and trying to explain the influence of b. The differences in the length spent on each parameter might be due to the students' own perception of the difficulty in the influence detection. So Iwan and Tom might have perceived b as the most complex and thus spent the most time to investigate this influence. They use the given technology to investigate the influences, however, as the sliders are difficult to set to specific values, they change them a number of times through manipulating the settings of the sliders.

The video analysis also shows that Tom and Iwan already identify all three influences correctly, while only working on part one of the worksheet, so describing the differences between $f(x) = x^2$ and $f(x) = 4 \cdot (x - 2)^2 + 3$. While moving the sliders to the values of the second example, they start investigating the influences and work on this investigation for ten minutes before reading the second part of the worksheet (Phase 1 to 9 in Figure 12.5). While Iwan describes the changes in the graphs, the two boys discuss if the transformed parabola has a y-intercept. During this, Tom uses the technology to convince Iwan that it does indeed have a y-intercept (see subsection 12.2.3). After ten minutes they proceed to part two of the worksheet and consolidate their findings by checking them again.

A common use of the technology in their process is the use of technology and then stating insight (e.g. Phase 6 and 11), which can be identified for example through analysis of the symbol sequences in the timeline (Figure 12.5). For example they change the slider for c and then b and state the influence of these two parameters. Tom also uses the technology to quickly check their insights again before writing them down, so for example when writing the influence of b and c, he checks if b and c have to be non-zero in order for the graph not to pass through the origin of the coordinate grid, or if it is sufficient if one of the two is non-zero (Phase 5 and 6). The technology enables to quickly visualize the possible values and therefore check the hypotheses before writing them down.

Iwan and Tom not only identify the influence, but also try to explain the changes through the parameter, for example for the influence of a they use the function table to explain, why $f(x) = 5x^2$ is narrower than $f(x) = x^2$. For this, they focus on single points $(1, 1)$ and $(1, 5)$ in the table and compare these two points (Phase 21). For parameter b, they also use the Grundvorstellung mapping to comprehend one of the given explanations on the worksheet (namely *if I look at* $(x - 1)^2$ *instead of* x^2 *all values in the table move towards bigger* x, see Phase 25). So the students are

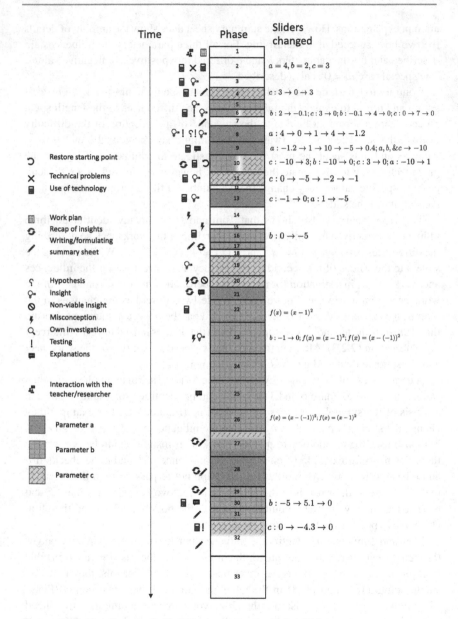

Figure 12.5 Timeline of a Guided Discovery Process Using Sliders

able to use their knowledge about function tables in an attempt to explain the newly discovered influence. These early explanations of the influence, for the up- and downwards movement through change of parameter c, the left and right movement through change of b and the vertical shrink and stretch through change of a using mostly single points in the function tables and the Grundvorstellung mapping are scarce in the videos (see also subsections 12.2.4 and 12.3.4). It might be that due to Iwan and Tom being rather high achievers, they succeed in their attempts to explain more than the middle to high achievers in the other groups videographed. But even for Tom and Iwan explaining the horizontal transformation other than with the given explanations is probably too difficult, as they do not succeed in explaining the influence. The horizontal transformation has been described in the literature as more difficult than a vertical transformation (e.g. by EISENBERG & DREYFUS 1994, BAKER et al. 2000, see subsection 2.2.2), so it is not surprising that explaining this transformation is also more difficult than explaining a vertical transformation.

During the 50 minutes of their Guided Discovery Tom and Iwan identify all influences correctly, attempt and even succeed in giving some explanations and are able to design a specific informative summary sheet. The structure given through the sliders and the worksheet seem to be beneficial for their investigation, especially the potential to quickly test and re-check their hypotheses. This second example of a Guided Discovery shows that the learning pathways of the students differ between different groups. In the next subsection the common points as well as the differences between the two learning pathways will be discussed.

12.1.3   Conclusion of the Two Examples

The two examples show, how different the Guided Discovery processes can be, but also reveal several similarities which are visualized in the timelines in Figures 12.2 and 12.5.

Both pairs worked mostly on their own for over 50 minutes and in both cases the Guided Discovery succeeded in evoking viable insights, but the way in which the viable insights are gained are different. While Charlotte and Noah followed the intervention task rather closely and gained insights in part two of the worksheet, Tom and Iwan identified the influence of the three parameters correctly, while still working on part one of the worksheet. So it could be reasoned that Iwan and Tom followed more the general task of exploration into the influence in the first ten minutes through their own curiosity. Charlotte and Noah, in comparison, might have needed the provided guidance to reach the goal.

The Guided Discovery approach enables the students to individualize their learning pathways through providing the freedom to deviate from the given tasks and explore on their own. This is for example visible, when looking at the two timelines of the two processes, as it can be seen that Charlotte and Noah work longer on parameter a, while Iwan and Tom spent more time on investigating parameter b.

In both processes, there are a number of aspects that only become visible through analysis of the video and are not visible on the summary sheets the students produced. If one only looks at the summary sheets, one might deduce that no explanations were attempted, however, Tom and Iwan achieve some degree of explanations on their own using the function table. Charlotte and Noah are encouraged by the researcher to attempt explanations, but they do not succeed.

The students are also even able to overcome some misconceptions on their own, for example the misconception that the graph is narrower if parameter c is changed (as described by e.g. GOLDENBERG 1988, PINKERNELL 2015 and subsection 2.2.3). This misconception occurs in both processes, though Tom and Iwan are only slightly confused by the statements regarding the misconception on the worksheet and overcome it immediately. Charlotte and Noah take longer to overcome the misconception and do so with the help of technology.

The two examples show similar uses of technology, the students use the technology to gain insights and test their hypotheses. The sequence of using technology and through this gaining some insight can be found a number of times. The possibility to quickly visualize a number of examples and check the conjectures seems to be beneficial, while the technological obstacles that both groups encountered were resolved rather quickly and did not hinder the insight completely. Another common use of technology in the two groups is that one student explains something to their partner using the technology (see subsection 12.2.3 for analysis of this use).

In conclusion, the analysis of the videographed processes allowed insight into different potentials and constraints of the approach that support or hinder the learning. Some of these potentials and constraints of the approach from the analysis of all 13 videos will be presented in more detail and as case studies in the following sections.

12.2   Potentials of the Approach

The videos showed a number of potentials of the chosen approach. The examples in this section are taken from a total of six different videos, but each described potential was visible in a number of videos. The examples were chosen to illustrate them. The identified potentials are both based on the technology and the Guided Discovery itself and were identified using the method described in section 8.4.

12.2.1 Affordance Through Use of Technology

The analysis of the solution summaries and transcripts led to identification of a number of affordances through the use of technology. Three aspects that were visible over a number of videos will be presented here using examples that illustrate these affordances: using technology to test hypotheses, using technology to explore on their own, using technology as a visual memory aid.

Using Technology to Test Hypotheses

Students in the experimental groups are using the technology to generate hypotheses, test them and generate insights through this testing. This corresponds closely to one of the uses of technology by experimental mathematicians described e.g. by BORWEIN (2005). Some students only test one small hypothesis quickly or double check their thoughts, while others combine a number of hypotheses and the testing of these to generate insights and thus go through a number of cycles regarding hypothesis testing. These sequences occur in a number of videos, an example of a short sequence using the technology to input one of the examples given on the worksheet, generating a hypothesis and testing can be found in one of the videos of the *function plotter* groups (see Figure 12.6). It shows a potential of the technology as it enables the students to test their conjecture through the quick visualization.

The two girls, Maria and Carla were working on part two of the intervention and investigating the influence of c. They already plotted the graph of $f(x) = x^2 + 5$ and Carla conjectures that the graph has its y-intercept at 5:

210. Maria: one, two, one, two, three, four (*counts the dashes on the y-axis*), yes it starts at five
211. Carla: one, two, three, four five (*also counts the dashes on the y-axis*), so it is that, where it begins, right?
212. Maria: yes, let's do minus one, then it should start at minus one

Line 212 shows that Maria conjectured correctly that the graph of $f(x) = x^2 + (-1)$ has a y-intercept of -1. The girls used the technology to confirm their hypothesis and thus gain a viable insight that c is the y-intercept in the functions $f(x) = x^2 + c$. This fast checking of conjectures enables the two students to quickly establish that they are correct. Sketching the graph for the function $f(x) = x^2 - 1$ per hand and

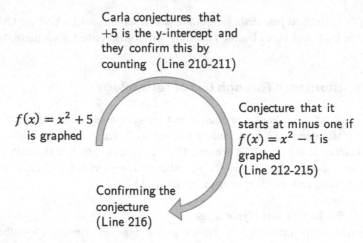

Figure 12.6 Checking Conjecture with Technology

then testing their hypothesis would have also worked, but taken longer and thus students are not able to sketch many examples. The fast visualization that is enabled by the technology can be seen as beneficial for this testing (see also section 3.2).

Another example of using the technology to check hypotheses is the following extract from two girls Inga and Tine working with sliders.[3] They do not only use the technology to check one conjecture at the time, but rather to check a conjecture, generate a new one out of the results of this checking and then testing the new conjecture (see Figure 12.7).

While working on task 2.1 they had set both sliders for a and b to zero and tried moving the slider for c to 5, but only achieved setting it to 4.9. Due to parameter a being zero, only a horizontal constant function is shown. The girls moved the slider for parameter a to zero as they wished to only change parameter c and did not realise that as parameter a is multiplied with x^2 they need to set the slider to 1 for the standard parabola. This problem occurs in a number of videos (see subsection 12.3.2 for discussion of this problem). Tine and Inga are surprised by this graph and search for the parabola:

110.   Tine: yes doesn't matter, wait, we can see (*zooms into the coordinate plane*), where is it (*moves coordinate plane*)
111.   Inga: uh it's gone

---

[3] This example has been also described in a shorter way in GÖBEL et al. 2017 and GÖBEL 2018.

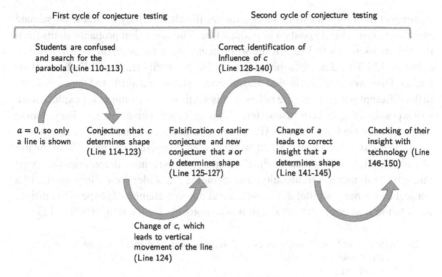

Figure 12.7 Sequence of Checking Conjectures with Technology

112. Tine: huh
113. Inga: huh that is only like this (*moves horizontally along the graph*)

Inga realises that the graph is only a horizontal line (in line 113) and not a parabola anymore. They then conjecture that *c* determines the shape:

114. Tine: If one *c*
115. Inga: (*snaps her fingers*) through *c* it becomes a parabola (*draws a parabola shape into the air*)
116. Tine: Right, wait let's do, ah no we can't close it
117. Inga: because (*moves coordinate plane*)
118. Tine: mhm
119. Inga: because now it is only a function like this (*follows the constant function with her hand*)
120. Tine: (*moves the coordinate plane*) Oh didn't think that, ha but wait ahm, four point 9 (*points with pen on slider of c*) ah minus yes
121. Inga: yes so when you, through *c*
122. Tine: through *c*
123. Inga: it is determined if it is a parabola, is the function
124. Tine: a parabola, yes ok (*Inga changes slider for c to 4.7*) and *c* also determines if they are far apart now the parabola or opened downwards or upwards right? Yes now it just moves

To support their hypothesis that c determines the change, they move the coordinate plane and argue that it is only a horizontal line. The confusion probably stems from the girls working on task 2.1 of the worksheet, which focusses on the influence of c. In line 124 Tine describes the influence of a, but attributes it to c. Earlier in their Guided Discovery they identified the influences viably and attributed them correctly to the different parameters (see below in this section as an example for exploration), so it is possible that this confusion stems from their earlier investigation. They change the slider for c (in line 124) and Tine observes that the constant linear function only moves vertically. Inga then correctly retracts their conjecture in Line 125. So the use of the technology helps them to falsify their hypothesis that c determines the shape through the dynamic visualization and changing the slider for c, only results in a vertical movement parallel to the x-axis and not in a change of shape. This insight leads to the new conjecture that a or b is responsible for the shape (in line 127).

125. Inga: no wait, wait, no, c doesn't determine that, that must be something else, because otherwise it would have changed
126. Tine: correct
127. Inga: Then it has to be one of the other, so a or b
128. Tine: so c only determines (*points to the graph*)
129. Inga: the y-intercept
130. Tine: but when it is not on the y-axis
131. Inga: ah yes
132. Tine: the
133. Inga: but but but but yes look the y-axis determines the vertex point on the y-axis, look here
134. Tine: but there is no vertex point
135. Inga: yes this one (*points onto the worksheet*)
136. Tine: But it doesn't intersect with the y-axis
137. Inga: yes, vertex point is vertex point, the vertex point is just the x-intercept down there
138. Tine: so c determines the vertex point
139. Inga: I would say so, depending on what happens now, (*changes slider for c to zero*) okay wait. Oah ha zero
140. Tine: yes then it is still

After falsifying their conjecture that c determines the shape, and stating their new conjecture that a or b is responsible for the shape of the parabola (in line 127), Inga and Tine also discuss the correct influence of c (lines 128–140). While Inga thinks c is the y-intercept, Tine realises that the vertex point of the parabola is not always on the y-axis. Together, they reach the insight that c influences the vertex point of the parabola. Even though no parabola is seen on the screen during this discussion as $a = 0$, the girls are able to connect the vertical movement of the constant function with the parabola. Tine remarks that there is no vertex point shown on the screen, but Inga realises that both graphs on the worksheet have a vertex point. This discussion might also have been influenced by their earlier exploration into the influence (as described below). After setting the slider for c back to zero (in line 139), they then proceed to changing the slider for a:

141. Inga: now comes a (*moves slider for a between* -2 *and* 9)
142. Tine: a? that determines it, so a determines if it is a graph or not
143. Inga: if it is a parabola or not
144. Tine: Parabola is what I mean, yes
145. Inga: okay, write that down

As soon as they move the slider for a (in line 141), Tine realises that a determines the shape of the parabola. So the technology helps the girls to verify their conjecture that a is responsible for the shape, especially the dynamic visualization of moving the slider and observing the corresponding change in the graph. Before writing it down, Tine wants to check this insight again using the technology:

146. Tine: wait when you change it to zero (*Inga tries to move the slider for a to zero*)
147. Inga: wait
148. Tine: yes…
149. Inga: there it is
150. Tine: yes
151. Inga: yes a determines if it is a parabola or not

So they check the verification of their conjecture using the technology by changing parameter a to zero, which results in a display of a straight line and they take this as a verification of their conjecture. So Tine and Inga use the technology to find out, which of the two parameters a or b determines if it is a linear function or a parabola. They also then recheck their result that a is responsible for the shape, before writing the insight down.

The above described sequence of using the technology to generate a hypothesis, falsifying this, generating a new one and then verifying this hypothesis is an example of the non-linear way the Guided Discovery can take, while reaching viable insights. The technology supports this sequence through the possibilities to quickly visualize the examples needed to test the hypotheses, as well as providing linked representation that has been shown to foster mathematical learning (see section 3.2). Both the short testing of one conjecture by Carla and Maria and the longer process by Tine and Inga show this potential, both pairs might have chosen to test the same things if no technology were available, but this would have taken considerably longer and and offered possibilities for mistakes, which would have led to wrong insights.

Using the Technology to Explore

Students are using the technology as described above to test specific hypotheses, and also to explore on their own and deviate from the given series of tasks on the worksheet. This was not forbidden or encouraged in the intervention, as investigation on their own might also lead students to the insights regarding the influences. The guidance was implemented, so if students did not know how to investigate on their own, they were able to follow some structure. But the free exploration without the guidance of the worksheet also leads to viable insight into the parameter influence in some cases.

An example for how the exploration can lead to viable insights, is again the process of Tine and Inga in the *sliders* group.[4] This excerpt is taken from the start of their process. After discussion about the function equation for about two minutes, while reading part one of the worksheets, the girls start exploring on their own and turn to the technology for the first time.

37. Tine: Yes, so but (*points with pen on equation on iPad*) for what is this formula there?
38. Inga: or, or, yes it is this, it is this
39. Tine: Yes it is this general
40. Inga: yes it is, here yes it is the general

They are identifying the formula $f(x) = a \cdot (x - b)^2 + c$ displayed in the app as the general function that is also given on the worksheet. They start with a in the equation and identify the starting value of a set by the slider.

41. Tine: now we look where a is (*points with pen to a in the equation on iPad*), a is 1 here
42. Inga: (*moves indexfinger on slider for a*) ok
43. Tine: What can you do with it? (*Inga moves the slider*)
44. Inga: different values (*moves slider to ca. 3*)
45. Tine: no bigger and smaller or so this or this (*opens and closes her fingers, see Figure 12.8*)

After moving the slider for a out of curiosity what one does with it (in line 43), they describe that the graph is smaller or wider and Tine makes a gesture representing the

---

[4]This example has been also described in a shorter way in 2017 and 2018.

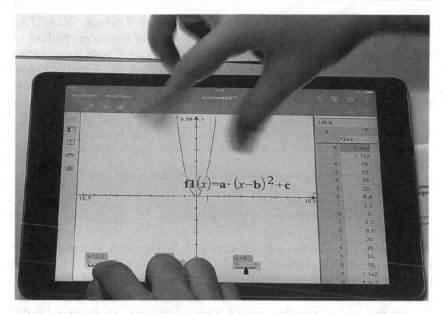

Figure 12.8 Screenshot of the Closing Gesture in Line 45

change of the graph. So they realise that *a* influences the shape and not the position of the parabola. The gesture (in line 45, see Figure 12.8) can be interpreted as an understanding of the vertical stretch and shrink through *a*. They then move to the slider for *b* and manipulate this (in line 48).

48. Inga: With b we go to (*moves finger to slider for b, manipulates it to three, but experiences problems with it*)
49. Tine: *b* what do you do with *b*, *b* moves it to right or left (*makes a gesture with her finger*)

Tine quickly realises that *b* influences the horizontal position of the parabola (in line 49), even though they experience some difficulty moving the sliders due to the size. The direct linking between the slider movement and the movement of the graph seems to foster this insight and enables the girls to gain this insight quickly. Finally they advance to explore the influence of parameter *c*.

51. Tine: exactly, there four yes like this and then (*Inga moves to slider for c*) with this you move it up, yes
52. Inga: four yes ok

While Inga moves the slider for c, Tine describes the vertical influence of c (in line 51). As they have set the slider for b to 4, the girls avoid the misconception that c is the y-intercept and instead only describe the correct influence of vertical movement. Again the fast visualization helps the girls to achieve their insight. Closing the period of exploration they recap their insights in order to write some notes for their summary sheet.

55. Tine: so a describes (*points with pen to slider bar for* a) this open and close, how wide it is open and how far it is closed (*Inga moves finger to slider b, then onto the equation*)
56. Tina: b describes? uhm right or left (*moves pen to right and left, while Inga moves slider for b*)
57. Inga: mhm exactly … exactly
58. Tina: c describes up or down, so move up. (…)

Tine recaps their insights for each parameter, while Inga moves the slider for b again, possibly to check their insight. In line 55 Tine also describes the connection between the value of a and the amount the parabola is stretched or shrunk. The girls did not, however, move the slider for a into the negative range and thus had no insight into the reflection through a in this sequence. Also, the slider for c is only moved in the positive range, but Tine still generalised in line 58 that the graph moves up- and downwards. The movement of the complete graph focusses on the Grundvorstellung object as a whole. The complete exploration described here only took one minute and fifty-three seconds, so in a very short time the girls explore the influence of the three parameters and gain some viable insights into the parameter influence. In this exploration the girls only use the three sliders as a kind of guidance and through this gain viable insights. So the curiosity of the students combined with the affordances through the technology, namely the dynamic linking between sliders and the graph and through this a quick visualization of the explored examples leads to a number of insights.

Using the Technology as a Visual Memory Aid[5]

Some students in the experimental groups not only use the technology to test hypotheses and explore, but also use the technology as visual memory aids, when recapping their insights for the summary sheet. In the *function plotter* group the students could graph as many functions as they would like and even turn the graphs on and off. The two students Carla and Maria in the *function plotter* group used this

---

[5]This aspect was analysed in collaboration with a master's thesis candidate, see HÖNEKE (2019).

extensively and did not delete most of the functions input, while working on part two of the intervention. When they start working on task three, Carla decides to turn the graphs of the functions plotted for the corresponding parameter back on, thus using the technology as a memory aid.

392.   Carla: yes then we briefly display again the ones we input for task, or? Uhm (*displays entry line*)

Carla has the idea to display the functions they worked on earlier, but the first time she decides to do this for parameter c (in line 392), fails, as they had earlier deleted the functions used for task 2.1. They then plot those functions again to restate their insights. Even though the first idea of just displaying the used function fails in this sequence, they still use the technology as a visual memory aid through plotting the same examples again. A successful attempt at re-displaying the graphed functions is evident while working on part 3.2, so explaining the influence of a. Carla realises (in line 442) that for explaining it, they should compare the graphs with the standard parabola.

440.   Carla: ok so
441.   Maria: (*reads out aloud*) explain the influence of a
442.   Carla: that's easier here you can compare it, compare it with the standard parabola, you can immediately so ay
443.   Maria: what was a again?
444.   Carla: wait (*displays entry line*)
445.   Maria: I hope it's not gone as well

Maria needs a reminder what a was and Carla displays the entry line in order to search for the relevant functions. Due to their earlier disappointment that the relevant functions were deleted, Maria states her hope to find the functions. So the girls have realised the advantage of being able to re-display functions they discussed earlier for their work. Carla succeeds in identifying the relevant functions they graphed earlier (in line 452).

448.   Carla: hello no (*turns the graphs of functions of*)
449.   Maria: but a times something
450.   Carla: that was up here somewhere (*scrolls through the turned-off graphs*)
451.   Maria: times yes that was
452.   Carla: yes, yes wait I'm looking where it starts, I think it starts here (*still scrolling through the entry line functions*)
453.   Maria: yes it is

454. Carla: you and you, thank you, no no *(displays functions for a* = 0, 3, 5) no go
      away, no, yes that was, how wide they were.

After Carla redisplays the functions for $a = 0$, $a = 3$ and $a = 5$, she recaps that a
influences the width and answering Maria's question (posed in line 443) what a was.
Carla also overcomes slight technological problems, as the app does not react the
way she wants, but she works around it. Even though using graphs as a memory aid
would also work with sketched graphs on paper, the amount of examples possible to
display and hide again, is a big affordance of the technology. The offloading of the
procedural parts (see section 3.2) of sketching many graphs also enables the girls
to display many functions at the same time in the same coordinate grid, but also
choosing only the necessary functions for a direct comparison.

   All three kinds of affordances using the technology present in the videos show that
students use the technology as an instrument (in the sense of instrumental genesis
described e.g by DRIJVERS & TROUCHE 2008, see section 3.1) as they use the artifact
aimed with some mental scheme for the use of it. An example for an instrumented
action scheme is the sequence by Carla and Maria for checking their conjecture (see
Figure 12.6). The affordance of using technology to test hypotheses matches the use
of technology "to test and especially falsify conjecture" (SINCLAIR 2004, p. 235) that
is used in a process of experimental mathematicians, so the students in this study
proceed to some degree in the same way as experimental mathematicians might
do while working to prove a theorem. The second affordance described above so
students using the technology to explore on their own could be seen as a combination
of the three uses "to gain insight and intuition", "to produce graphical displays that
can suggest underlying mathematical patterns", and "to discover new patterns and
relationships" (SINCLAIR 2004, p. 235) as the students gain insight through their
exploration of new examples, so this also represents to some degree the same way
of working by the students as experimental mathematicians.

12.2.2 Drag Mode Leads to Different Direction of Change of Representation

Change of representation is important for conceptualization as described above (see
section 2.4). The direction of the change in the intervention was mostly from the
algebraic term to the graph, so the students would sometimes argue along the line
"if you change a, the graph is wider". However, some students in the *drag mode*
group argued in the opposite direction so along the line "if you drag the graph at this
point, a changes". This is a possible starting point to foster the bidirectional change

of representation between graph and equation. An example how arguing this other direction of change of representation is a sequence of Nina and Rose working with the drag mode. They tried inputting $b = -5$, but are confused with the simplification of the function equation that is shown. Nina then argues with moving the parabola into the positive area of the x-axis so the sign in the function equation is a minus.

145. Nina: it moves into the positive area, rather move it there (*points to the right*)

Arguing this direction as Nina does in line 145 might be seen as non-viable, but the possibility of the drag mode combined with emphasis on the change of representations of graph to algebraic term could offer potential to foster the linking of these two representations, which has been described as crucial for learning (see e.g KAPUT 1992 and section 2.4). This direction of arguing the influence circumvents the necessity of operating with parameters to some degree as direct manipulation of the graph results in a numerical change in the function equation in contrast to the sliders group, where students manipulate a slider titled with a parameter and observe the change of the graph. This direct manipulation of the graph therefore could help students with an incomplete understanding of parameters achieve some kind of insight into transformation of functions, but at the same time the connection between the parameter and its influence on the graph is somewhat obscured.

12.2.3 Communication With and of Technology

The students worked in pairs during the intervention and due to this there necessarily occurs communication between them. Using the framework developed by BALL & BARZEL (2018), examples of using the technology to explain something to each other and the technology screen as a stimulus for discussion were identified in the videos. The technology screen as a stimulus for discussion was described by BALL & BARZEL (2018) as communication of technology and they propose these stimuli as a kind of scaffolding for learning (BALL & BARZEL 2018). With this the students do not interact with the technology, but only refer to the display and discuss what is shown. Using the technology to explain something to each other, encompasses both the communication of technology and the communication with technology described by BALL & BARZEL (2018) and will be called communication including technology. With this the students interact with the technology (which is described as communication with technology) and also use the technology screen as a stimulus for discussion or interaction between them. Both kinds occur in all three experimental groups and examples will be presented using transcripts.

Communication of Technology

An example for using the technology display as a stimulus for discussion and discussing about what was shown by the technology is taken from two girls, Nina and Rose working with the drag mode. In the example they are trying to comprehend the second statement given for explaining the influence of b ("$x^2 = 9$ for $x = 3$, but $(x - 1)^2 = 9$ first applies for $x = 4$.") in part three of the worksheet and moved the parabola to $f(x) = (x - 1)^2$. They identify the parabola on the screen as the graph of the function given in that statement and Nina then tries to retrace the statement:

298. Nina: it was moved to the (*points to the positive part of the x-axis*) here you see that, I did not zoom in, but (*makes a zoom gesture*) made it bigger uhm, here somewhere is four, there is four (*points to x = 4, see Figure 12.9*) and I would say, here is nine (*points to y = 9*) and that is correct, because that is the point and normally it would be at three and nine.

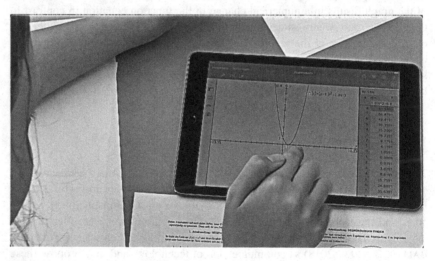

Figure 12.9 Screenshot of the Pointing Gesture in Line 298

299. Rose: yes that is correct. Yes uhm

In line 298 Nina uses the graphed equation and the given statement to comprehend the statement using the technology as a visual aid as well as a stimulus to discuss the points without interacting with the technology. She uses the Grundvorstellung mapping and identifies for $x = 4$ the corresponding y-value (in this case 9) on the graph. This is part of the statement in the worksheet, which also uses the Grund-vorstellung mapping and comparison of two points on the two graphs $f(x) = x^2$ and $f(x) = (x - 1)^2$ as explanation. So Nina comprehends part of the statement using the technology display as a visual aid. The *communication of* could also occur with printed or sketched graphs, but the technology allows for a wider range of examples as well as the affordance of interacting with it, which leads to the *communication including* technology examples.

Communication Including Technology

Noah (see Phase 25 in Figure 12.2 and section F.1) uses the technology to explain, why the two of them made a mistake earlier in the process. While writing their summary sheet, he goes back to their statement that the graph is narrower, if c is changed and wants to check this. He plots the functions $f(x) = x^2 + (-5)$ and $f(x) = x^2 + 5$. Charlotte states that she does not know why the graph is narrower, even though it seems narrower. Noah explains (in line 369) why this assumption is wrong:

369. Noah: it does not get narrower, Charlotte it only runs differently, have a look here, (*Noah moves coordinate plane and zooms in, see Figure 12.10*) in the end it is just as fat.

Figure 12.10 Screenshot of the Movement in Line 369

370. Charlotte: agreed.
371. Noah: only because it just starts a bigger curve, because it starts further down (*Noah draws a parabola shape with his hands in the air*). It is exactly the same graph.
372. Charlotte: yes then we have this, but then it is not Sonja, but him, he is right then
373. Noah: yes it is exactly the same graph only, because it only starts further down or further up

Noah uses the technology to test their earlier insight and graphs the two extremes they tried for parameter c in task 2.1. Charlotte tries to explain, why the graph gets narrower, however Noah uses the technology to explain that it does not get wider or narrower, but it is the exact same shape only moved upwards (in line 369). In this case, Noah uses the Grundvorstellung object as a whole to argue his viewpoint. He supports this by scrolling through the coordinate plane and explaining that the graph starts below the other, but the shape is the same and Charlotte admits that they made a mistake. Charlotte refers in Line 372 to the task 3.1, where they stated that Sonja gave the correct statement and corrects this. So the possibility to move the coordinate grid and therefore changing the window allowed Charlotte and Noah to overcome the common misconception that c influences the shape, which has been described in the literature for example by GOLDENBERG (1988). This flexible changing of the window would not be available, if students sketched the functions per hand. Also it provides a valuable tool for Noah to explain this mistake to Charlotte, who was earlier convinced that the graph of $f(x) = x^2 + 5$ is narrower, as he can show his argument to her by moving the coordinate plane and thus supporting his verbal correction.

Another example for the use of communication including technology can be found in the process of Tom and Iwan (see Phase 3 in Figure 12.5 and section F.2) very early in the Guided Discovery. They had input the example from part one into the app and set the values to $a = 4$, $b = 2$, $c = 3$ (see Figure 12.11).

The boys were describing the change between the graphs of the two functions $f(x) = x^2$ and $f(x) = 4 \cdot (x - 2)^2 + 3$ in part one of the worksheet. Even though both seem to work on their own, they listen to the description of the other one and respond accordingly. In line 26 Iwan states that there is no y-intercept, while pointing to the graph of $f(x) = 4 \cdot (x - 2)^2 + 3$ on the worksheet. Tom immediately responds that the graph has a y-intercept (in line 27).

25. Tom: is narrower. What? What?
26. Iwan: so firstly the graph changes, there is no y-intercept at all anymore
27. Tom: Indeed!
28. Iwan: in the graph it is not shown at least.

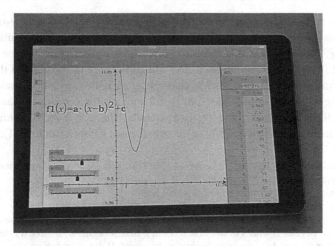

Figure 12.11 Screenshot of the Display in Line 25

29. Tom: there, (*begins to move the coordinate plane and zooms in in order to show the y-intercept*) it definitely has one.

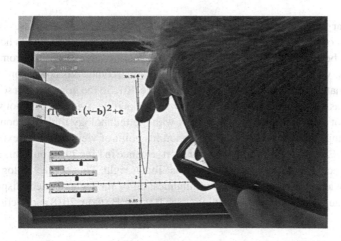

Figure 12.12 Screenshot of the Movement in Line 29

Iwan corrects that it is not shown in the graph (which is correct, as the y-intercept is not visible on the worksheet due to the window settings) and Tom turns to the technology to show Iwan that the graph indeed has a y-intercept (in line 29, see Figure 12.12) using the possibility to zoom in and change the position of the coordinate plane. They describe further changes again both working separately, but responding to each other. While Tom is still moving the coordinate plane, Iwan describes the vertex point of the standard parabola. This leads to some confusion as Tom assumes Iwan is still describing the transformed parabola (in line 31).

30. Iwan: The start, definitely it begins at zero. Zero zero (*he means the origin of the coordinate plane*)
31. Tom: (*still moving the coordinate plane*) No.
32. Iwan: huh? With the standard not? Here (*points to the worksheet*)
33. Tom: (*still moving the coordinate plane*) yes with the standard parabola, yes
34. Iwan: yes that is what I mean, here it doesn't, here it begins at, at three it is logical
35. Tom: (*the app does not react in the way he wants*) Hello, stupid thing
36. Iwan: Uh, but there is something not right, here c should be at three, oh this is three, isn't it?
37. Tom: It is
38. Iwan: ah yes, ok
39. Tom: there, three
40. Iwan: now I see it uh

They clear this confusion and both agree that the standard parabola passes through the origin of the coordinate plane. Iwan suspects a mistake (in line 36) as he thinks c should be three. This confusion might stem from overgeneralising from linear functions, where c is the y-intercept. But Iwan overcomes this confusion that the y-coordinate of the vertex point is three (see line 40). So the graphic representation first acts as a stimulus for discussion as Iwan thinks if the y-intercept is not visible, it does not exist and then Tom interacts with the technology through zooming in and moving the coordinate plane to support his point of view that there exists a y-intercept. Even though he only speaks a short sentence in Line 29, through his actions with the technology, Iwan is convinced that Tom is right. Thus the technology leads to a knowledge gain through the communication of the technological displays as well as interaction with the technology (see BALL & BARZEL 2018 and section 3.5).

12.2.4 Early Versions of Explanations

Even though many students did not achieve in finding complete explanations for the influence of parameters, some students did achieve early versions of explanations. Some students used the Grundvorstellung mapping to explain the influence of a and compared two points on the graphs, so for example $(1, 1)$ for $a = 1$ and $(1, 5)$ for $a = 5$ using the function tables.

An example of these early explanations not only for parameter a, but also for c is the Guided Discovery process from two girls, Grace and Laura from the *function plotter* group. They quickly plot nearly all examples given in task 2.2 (influence of a) and describe the influence correctly. Visible on the display are the functions $f(x) = a \cdot x^2$ for $a = -5, -1, 0, 0.5, 1, 5$. Grace then wants to understand, why the shape is different and states:

173. Grace: that is wider, but now I do not know
174. Laura: yes we described it now
175. Grace: why it is. Ah yes, have a look, when it is at one it is, I get it, have a look, when it is at the thing one is, then it is here a half (*points to* $(1, 0.5)$ *on the graph* $f(x) = 0.5x^2$, *see Figure 12.13*), when it is five times x it is one (*points to one on the* x-*axis*) five (*points to* $f(1)$ *for* $f(x) = 5x^2$)
176. Laura: so it is the slope

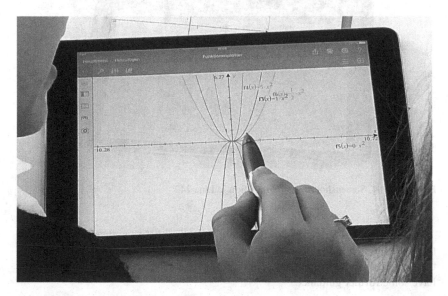

Figure 12.13 Screenshot of the Explanation in Line 175

Grace uses the Grundvorstellung mapping to compare the function values of $x = 1$ on the different graphs and at the same time connects this with the shape of the graph, however, Laura still identifies it with the slope, this is not correct as quadratic functions do not have a constant slope But as a first explanation of the influence it contains many viable points, as the slope is influenced through a and the approach by Grace to compare the different graphs point-wise is a very good attempt at explaining the vertical stretch and shrink.

The girls successfully use this kind of explanation also for the parameter c, after the researcher points them in the direction of using the function table. Grace does this and inputs the functions $f(x) = x^2$ and $f(x) = x^2 + 1$ and displays the corresponding function table (see Figure 12.14). Grace then identifies the corresponding row in the table:

242. Grace: uhm x squared plus one (*presses the input button*) so and now we have
       to display the function tables (*displays the function tables*) and now look, no it is
       always only plus one, one hundred, one hundred and one (*points to the function
       table*), one-hundred-twenty-one, one-hundred-twenty-two.
243. Laura: correct

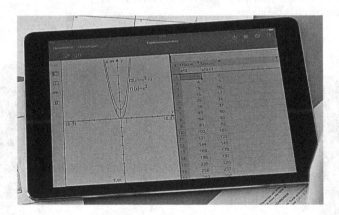

Figure 12.14 Screenshot of the Explanation in Line 242

Grace manages to identify the constant difference between the two columns in the function table, but does not generalize how this could be an explanation, why the graph does not change shape (in line 242). To achieve this they might need to have some further guidance.

They then start working on their summary sheet and once they reached their insights for parameter a, Grace wants to find explanations again using the function table, without realising that she already had some degree of explanations earlier in the process. For this she inputs the functions $f(x) = x^2$, $f(x) = 5x^2$ and $f(x) = 0.5x^2$ again and displays the function table.

279. Grace: say look here you can see that one, well, there is nothing here (*points to the equation of the standard parabola function table*), then it is one and here one times five is five (*points to the cell in the table for f2(1) with $f2(x) = 5x^2$*), one times zero point five and then here three times well three squared is simply nine (*points to $f1(3) = 9$ in the function table, see Figure 12.15*)

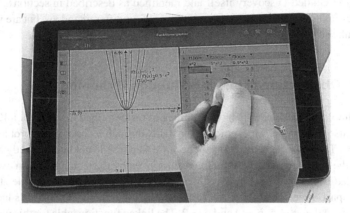

Figure 12.15 Screenshot of the Explanation in Line 279

280. Laura: yes

Again, Grace chooses a point-wise approach for the explanation and compares the two columns of function values line by line using the technology to display them and point to the corresponding values. So Grace succeeds both in explaining the influence of a using the graphs as well as the function table, both times using the Grundvorstellung mapping.

The guidance given through the worksheet and the hint from the researcher to use the function tables is enough for Grace to start with explaining the influences of a and c. In the final minutes of their process, Grace also uses the Grundvorstellung mapping to comprehend the left statement of part 3.3 of the worksheet (see Figure 7.18, "$x^2 = 9$ for $x = 3$, but $(x - 1)^2 = 9$ first applies for $x = 4$."). These early versions of explanations in their process are mostly viable and show the potential, however as these explanations are scarce over all processes and even less evident on the summary sheets (see subsection 11.1.7), more guidance should be implemented to foster more explanations.

12.3   Constraints of the Approach

The section above presented the potentials of the approach that were visible during the intervention. However, there were also a number of problems of the approach visible. As with the potentials the difficulties are both based on the technology as well as the Guided Discovery itself and identified as described in section 8.4. The examples in this section are taken from a total of four different videos, but are visible in a number of videos.

12.3.1 Obstruction of Insights Through Technology

The technology does not only have affordances (as described in subsection 12.2.1), but can also hinder the achievement of insights through technological problems or obstacles. The obstacles mostly occur, because students do not know exactly how to operate the software or to interpret the results given by the technology. An example for the latter can be found in the slider process of Inga and Tine. While starting work on part three of the worksheet (explaining the influences), the girls have set the sliders so that $a = 5, b = 4$ and $c = 2$. The linked function table for this function is also visible on the screen (see Figure 12.16), while on the worksheet the function table for the functions $f(x) = x^2$ and $f(x) = x^2 + 1$ is given (see Figure 7.18 for the given function table). The students turned their attention to the table and try to understand it:

224. Inga: and what is it with the function table here (*points to the function equation on the worksheet*)
225. Tine: ah so yes the function table?
226. Inga: but look there are two columns there (*points to the table on the worksheet*), here we only have one.
227. Tine: I, especially, I do not get what this *e* means, there is nothing there or? (*points to the display with a pen first to the scientific notation in the function table, then to the graph, see Figure 12.16*)

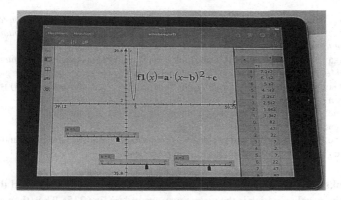

Figure 12.16 Screenshot of the Scientific Notation in Line 227

228. Inga: No

While Inga is confused, that the function table that is given on the worksheet for task 3.1 (explaining the influence of parameter c) has two columns compared to their function table with one, Tine is confused by the scientific notation in the function table of values above 99 (in line 227). Tine thinks that there is nothing in the graph for negative value. This problem probably stems from an unfamiliarity with the software and therefore inexperience with the scientific notation. They do not try to zoom out, which would lead to their assumption being disproved. Instead they continue trying to understand the function table:

229. Tine: So at two c
230. Inga: wait point
231. Tine: What point is it?
232. Inga: yes point four at two
233. Tine: like a coordinate plane, so two twenty-two is here, then (*points with pen to the line in the function table where $f(2)$ is*), two and somewhere there is twenty-two (*searches for $x = 2$ on the x-axis and then moves upwards to twenty-two on the y-axis*). So theoretically it is then the huh

They correctly suspect that the two columns give the coordinates of the parabola and Tine gives an example using the point (2, 22) and showing it on the graph (in line 233), so she uses the Grundvorstellung mapping to understand the function table. But a question from Inga evokes a mistake with Tine and she shortly confuses the x- and y- value in the table (in line 235):

234. Inga: But I still don't understand this x (*points to the header of the function table*)
235. Tine: That has to be, that has to be then the y-value (*points to the left column in the table*) and this x (*points to the right column in the function table*)
236. Inga: But? Wait, wait, wait, wait, look here, wait, it begins (*points to the function table*) there, with us it begins here at four (*points to the vertex point of the graph*) or something at here four (*moves her finger to the graph at around $x = 4$*) is two uhm
237. Tine: But then there has to be a point at two and four or something
238. Inga: two, yes
239. Tine: That means two and twenty-two (*points to the table*), yes there four and two, then that is the x-value (*points to the first column*)

Tine thinks that the left column in the function table is the y-value and the right column the x-value, which is the wrong way around (in line 235). They then pick one line out of the table and try to identify this point in the graph, while still confusing the x- and y-axis. In their search for the point on the graph in their table (in line 237), Tine realises their mistake and correctly identifies the first column in the table as the x-values (in line 239). So the function table linked with the graph enables the girls to connect the two representations as far as identifying the two representations as belonging to the same function. This is an important step in understanding as this is a first step in order to change between them (see DUVAL 2006 and section 2.4). Following this, they discuss about the—apparently unusual—notation using the e, which was already seen as unusual by Tine in line 227:

240. Inga: But then look here with the e it is, we have the parabola opening upwards not downwards (*moves her finger upwards along the y-axis*), thus maybe it is invalid or something
241. Tine: So yes I think it is only at minus 7 (*points to minus 7 on the x-axis*), when one is at three point two one minus one, nah minus one goes downwards then or?
242. Inga: uhm
243. Tine: here so it is on this side with the x-value (*points to the negative side of the x-axis*) and then it goes up and then there is no parabola
244. Tine: so it doesn't intersect at all, doesn't it?
245. Inga: but the parabola goes on and on with c, that means it has to reach the other side some time, doesn't it? Because otherwise
246. Tine: yes at zero and eighty-two it is like that (*points with her pen to the origin of the coordinate plane*)

They try to find reasons for the unusual notation, however, do not succeed in doing so. They conjecture that it might not have y-values for negative x-values, even though Inga correctly states that the parabola does not stop and should reach the other side. Tine thinks that the y-intercept suffices that condition. In this case, Inga shows that she has some understanding of functions regarding that they do not stop at a certain point, when she argues in line 245 that the parabola has to reach the negative side of the x-axis at some point due to the endlessness. It might also be using a Grundvorstellung object as a whole, as she shows some knowledge about the general run of a parabola. This understanding might not be robust enough yet, as Tine's argument in Line 246 that the parabola has a y-intercept at $(0, 82)$ seems to satisfy her. The y-intercept is not shown in the window, but they do not try to zoom out or move the coordinate plane, which would result in the y-intercept appearing in the window and therefore might have helped the girls overcome this obstacle. In this sequence it would have been beneficial for the teacher to be present and provide guidance to resolve this problem, then the students might have used the function table to explain this influence. The girls do not discuss this problem further, but return to the tasks on the worksheet.

So the unusual display of the values in the function table leads to confusion and hinders insights. But at the same time it can also be seen as an example of *communication of technology* (as described by BALL & BARZEL 2018), as the confusing notation leads to discussion and reflection about the function table. It can only be conjectured, what the girls would have done with the function table if the scientific notation was not used, but it might have led to reasoning using the table. In this case the technology used confused the students, this might not have happened if the students had used a different app or computed the function table by hand. Also the problem of the unusual notation might have been prevented if students had worked for a longer period of time with the technology and therefore had encountered and overcome this problem before. It could be argued that in this case the students had not developed deep enough usage schemes to use the technology as an instrument (in the sense of instrumental genesis, see e.g. DRIJVERS & TROUCHE 2008 and section 3.1).

Missing the technological knowledge how to operate the app at all can also lead to delay in insights. This is especially the case in the *function plotter* group, as they cannot just manipulate the graph through dragging either the graph or a slider, which might be seen as more intuitive by the students. Even though all students had an introduction into TI-Nspire CX CAS the students Noah and Charlotte encountered problems, when they first tried to use the technology to graph the function, while working on task 2.1 (influence of c). They wish to plot a function using the app.

49. Noah: say we, wait, where do we have to input it, so the things here yes there (*displays entry line*).
50. Charlotte: fx x to the power, fx equals x to the power

51. Noah: yes uh x squared (*taps the button on the keyboard that has x^2 written on it, this only enters an empty box to the power of two, as the variable has to be entered first, see Figure 12.17*)

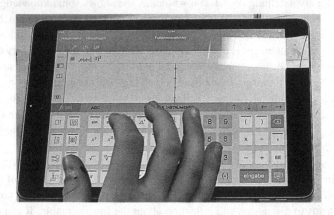

Figure 12.17 Screenshot of the Entered Empty Box to the Power of Two in Line 51

52. Charlotte: plus
53. Noah: huh, but there is
54. Charlotte: plus minus five
55. Noah: I want x squared (*taps repeatedly on the x^2 button*) no this was, no this was wrong okay so huh I do not get it, so
56. Charlotte: wait we had this weird info sheet, no we did not (*looks at the back page of the worksheet*) ah no that is that no.

Noah encounters the problem that if he taps the button, which is labeled x^2 it only enters an empty box to the power of two (in line 51). He does not realise that he might need to input the variable separately and repeatedly taps this button (line 55). In the introduction into the TI-Nspire App students were given an information sheet, which Charlotte searches for in line 56. In the introduction into the TI-Nspire App the students only graphed linear functions (see Figure B.1), so the problem of not knowing how to graph the quadratic function could have been prevented. As they did not achieve to resolve this issue on their own, they ask the researcher and she shows them how to enter x^2. They then succeed in entering the function equation, but come across another technological hurdle, as they do not know, how to plot the function (line 67):

67. Noah: close the parenthesis, no so uh why is it just gone (*closes entry line while trying to plot the graph*) and now? Nothing happens, nothing happens (*tries to display entry line again*) now it is gone, why is it gone? Why is it gone? (*Charlotte displays entry line again*) yes that gets huh

68. Charlotte: just input it again

69. Noah: yes ok here (*inputs the equation*) x to the power of two plus huh why then I can just input minus five.

70. Charlotte: Yes, I said so

71. Noah: minus five (*to a class mate*)so cool. No I still do plus minus five, that can stay like this, plus close parenthesis minus five, so have a look where do we do it now, I would huh enter (*taps by chance on the enter button*) so and now? (*reads aloud*) observe what happens, ah so now for the different values

72. Charlotte: exactly

The problem of not knowing how to plot the entered equation is solved by chance, as Noah presses the enter button by chance (line 71). The whole sequence from the first decision to plot a function to the function finally being plotted was about three minutes long, the technological problems delayed the achievements therefore. Even though three minutes might not be considered long, with a class length of 60 minutes it does have an effect. The technology therefore delays the insight. These problems could be prevented if the students would work with this kind of software more often and therefore develop more stable usage schemes. Thus, students would then be able to use the technology as instruments. But the usage schemes can develop quickly and some features like the drag mode are more intuitive than others, so the students still achieve to use the technology in an aimed way most of the times.

12.3.2 Value of a in Standard Parabola

In the intervention task, the students were asked to investigate the different parameters one at a time and many students did so in the order given on the worksheet so c, then a and finally b. For the structure the function equation $f(x) = a(x - b)^2 + c$ was reduced by setting two parameters that were not looked at in the subtask to zero or one depending on the parameter and then simplifying. So for the third task regarding parameter b, the function equation was simplified to $f(x) = (x - b)^2$. However, especially for the *sliders* group this led to the problem that students did not know what value they had to set for slider a.

The problem also occurred in the *drag mode* group, because as soon as they dragged the parabola, so that a changes, the function equation has a value in front of the x^2, whereas on the worksheet the function is given without the factor in front of the x^2. This leads to the discussion about the value of a in the standard parabola.

The two girls Nina and Rose working with the drag mode encounter this problem at various stages of their Guided Discovery and only resolve it at the end of their process. This example shows that the students' Guided Discovery does not progress in a linear way, but rather the girls tend to encounter problems or insights a number of times during their 60 minute long work phase.

The first time, they come across the problem, is while working on task 2.2, where they are supposed to change a:

79. Rose: ok uhm (*reads out aloud*) Now look at a. Use different values of a in $f(x) = a \cdot x^2$, for example minus five. Ok how do we change a? Is there even an a? (*see Figure 12.18*)

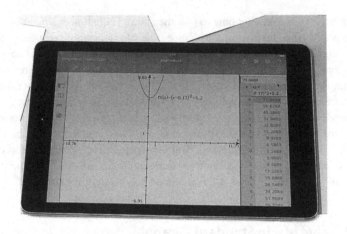

Figure 12.18 Screenshot of the Display in Line 80

80. Nina: no I don't think so
81. Rose: a
82. Nina: no there would, look here, there it is only squared so technically there is no a

In line 79 while the function $f(x) = (x - 0.17)^2 + 5.2$ is graphed (see Figure 12.18) Rose poses the question if the function in this case even has an a. They agree that it does not have an a. This problem of not realising that if a variable has no value in front of it, it is multiplicated by one, might stem from a not a robust enough variable understanding (see also subsection 2.3.3 for a discussion of misconceptions). Another reason for this might be that the basic convention of omitting multiplication by one in writing is not clear and therefore could cause the irritation. Later in the process it seems like the convention was taught however, as Rose states this convention and it is indeed a misconception, which is overcome later. They decide to move the parabola to the vertex point again, as this was the position the graph was at the beginning of the Guided Discovery and succeed in nearly doing so. They then achieve in changing the value of *a* through dragging (in line 88) and assume that only now *a* is present in the equation.

88. Rose: exactly won't work, but uhm so *a* is not here until now. Now it is only x minus b ok let's see if it changes now (*drags the parabola and changes a*) yes, now there comes *a* (*see Figure 12.19*)

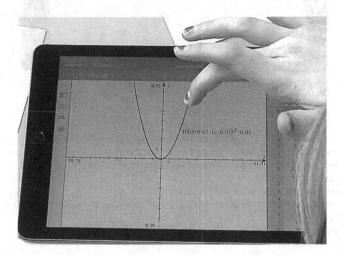

Figure 12.19 Screenshot of the Display in Line 88

The dragging at the parabola causes a change in parameter *a*, which is now displayed in the function equation. They do not discuss the problem of the value of *a* further, while working on task 2.2, but rather describe the course of the parabola for negative and positive values. They proceed to identify the influence of *a* including the reflection for negative values and the special case *a* = 0. They only encounter the problem of the value for *a* in the opposite direction again, while working on task 3.3 (explanations for *b*), when they want to remove the *a* in order to graph the standard parabola as asked in the worksheet.

212. Rose: x squared is technically exactly in the middle here (*see Figure 12.20*)

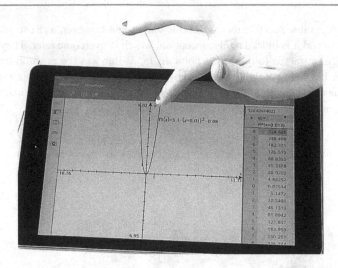

Figure 12.20 Screenshot of the Display in Line 212

213. Nina: yes, you have five point one (*the graph of* $f(x) = 5.1 \cdot (x + 0.01)^2 - 0.08$ *is shown*) here, so you have a though (*points to the graph*)
214. Rose: yes, that means actually, how do I remove a? (*changes a through dragging*) I want to get rid of a
215. Nina: yes, you have to go further down then.
216. Rose: yes, but I want (*incomprehensive, still dragging the parabola, stops at a =* 0.1)
217. Nina: huh but at zero, there is no slope then at all

In order to achieve the standard parabola the girls moved the parabola back to the origin of the coordinate grid, but Nina realises that they have a value of 5.1 for a and therefore Rose wants to remove a. Rose tries to remove a by shrinking the graph through dragging, however, Nina realises (in line 217) that if they keep going it would be a graph with a slope of zero. They stop at $a = 0.1$ and Rose thinks it is the standard parabola (in line 218), so confuses the value of a by a factor of ten.

218. Rose: that is actually the standard parabola
219. Nina: yes like this (*incomprehensive, points to the graph*), right?
220. Rose: yes, but it is then, wait
221. Nina: because the standard parabola, if this here are units, then it would always go through one one (*points to the graph and the dashes for* 1 *on both x- and y-axis*)

Nina falsifies their conjecture of line 218 that it is the standard parabola by testing the point $(1, 1)$, through which the standard parabola passes, but the parabola on their screen (which is the graph of $f(x) = 0.1 \cdot (x - 0.01)^2 - 0.08$) doesn't. She uses her pre-existing knowledge of the standard parabola combined with the Grundvorstellung mapping as she explicitly states that with the standard parabola the x-value of 1 is mapped to the y-value of 1 to falsify their conjecture. Knowing that it is not the standard parabola, they discuss to find out, what they have to change. The girls agree that the minute translation of the graph in the vertical direction is not a problem (in this situation $c = 0.008$), as they can ignore this and knowing that due to the accuracy of the drag mode will not succeed in removing it (line 227). In this discussion, still revolving around removing the numerical value of a in front of the equation, the girls use the technology to try and achieve this:

222.   Rose: yes exactly. So uhm what do we have to change there?
223.   Nina: so I would say here *(points to the display)* this minus c is not so bad, because we will not get rid of it, so at the very back
224.   Rose: yes, exactly and we really only have to get rid of a, the rest is so *(Nina taps onto the graph)*, but
225.   Nina: *(drags the graph so it shrinks further)* no, if I do that further, it is flat again
226.   Rose: yes, exactly
227.   Nina: or?
228.   Rose: yes, actually already *(Nina drags the graph, so a gets smaller)*
229.   Nina: ok, this is weird *(changes a, so it gets bigger)*
230.   Rose: *(incomprehensive)* achieve the standard parabola? We had it at the very beginning, we had it then, right? And it looks like this *(points to the worksheet)* But go back to the beginning? *(presses the undo button)*
231.   Nina: try it

As their attempt to remove a from the function equation fails, they decide to undo their changes until they reach the point (in line 230), where no a is displayed in the function equation. This decision stems from Rose remembering that the standard parabola was displayed at the beginning, so she monitors their progress of the Guided Discovery in some kind and decides to take a step back to gain further insight.

238.   Rose: uhm yes *(repeatedly presses the undo button)* wait, actually we should *(incomprehensive)* no no
239.   Nina: yes now we should *(incomprehensive)* everything negative, that is it actually, right? *(points to the display)* there we did not have an a yet.
240.   Nina: yes there we did not have an a, wait, I'll drag it downwards *(moves the parabola downwards)* I still do not know why

Achieving to undo their changes until the parameter a is not displayed in the function equation, the girls move the given parabola back to the origin (in line 240), without achieving insight into the problem of the value of a. They then return to the tasks on their worksheet and proceed in their Guided Discovery, but the question of the value of a remains. This is resolved at the end of their process, after they finished their summary sheet and have some time left. Nina then decides to investigate the question again.

528. Nina: (…) I still do not understand, why the standard parabola has no a
529. Rose: I do not understand it as well
530. Nina: can I try something? We still have time anyway (*starts dragging the parabola towards the origin of the coordinate plane*)
531. Rose: yes, just

In order to investigate the problem, why the value of a is not always displayed, Nina wants to move the parabola to the origin, which she does not achieve to do so, due to the accuracy of the drag mode. She works around this problem by pressing the undo button, until the parabola is nearly in the origin. They then discuss the reason for the missing a.

542. Nina: ok, so I do not understand, if there is actually one times something, you can omit the one (*see Figure 12.21*)

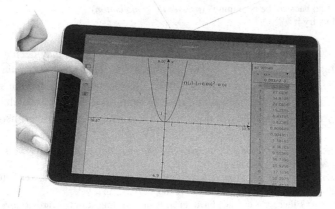

Figure 12.21 Screenshot of the Display in Line 542

543. Rose: yes
544. Nina: now I'm thinking if there is one times x plus something to the power of two. No (*has a problem choosing the parabola and then changes a*) now it is zero point
545. Rose: ah so x is probably
546. Nina: but then it should be written there or not? (*still dragging the parabola*)
547. Rose: yes, one time

Nina starts her investigation with the help of the technology and her hypothesis based on previous knowledge of the convention that one is the identity element of multiplication and therefore can be omitted (in line 542). She then assumes that this might be the case in this problem as well, so she transfers her previous knowledge onto the new investigation. To test this she changes the parabola through dragging (in line 544) for values $a < 1$ and returns the graph to $a = 1$. Due to the accuracy and rounding now the value of a is displayed as 1 and Nina identifies it (in line 548) as the standard parabola:

548. Nina: wasn't it like this? (*pointing to the display shown in Figure 12.22*)

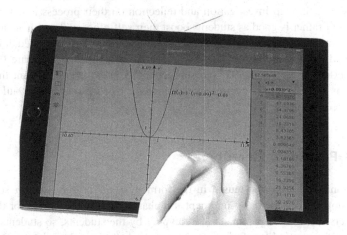

Figure 12.22 Screenshot of the Display in Line 548

549. Rose: yes one times x is exactly the standard parabola
550. Nina: it was like this (*presses the undo button once, and the value of a in the equation vanishes*)
551. Rose: yes, exactly that is exactly one
552. Nina: yes, that is why it was omitted. I just wanted to clear up, why it was left out.

In order to confirm that $f(x) = 1 \cdot x^2$ is indeed the standard parabola, Nina presses the undo button and then confirms that the parabola does not change shape and only the value of a vanishes in the function equation. She then concludes that the 1 was only omitted and therefore no value of a was shown in the equation.

This example shows how the progress in the implemented Guided Discovery is not linear, but the students tend to revisit and re-encounter problems during their investigation. The problem of the value of a in the standard parabola is a significant obstacle in their course and might lead to obstacles in the Guided Discovery. This should be addressed in class, as it is an important point in the conceptualization of variables and parameters to learn, when numerical values can be omitted and when they have to be kept. To avoid this confusion in the Guided Discovery, the value of each parameter could always be written in the function equation even if it is 1 in the case of a or 0 in the case of b and c or the students reminded of the convention of omitting the 1. Thus, this confusion might have been prevented and the considerable amount of time, the two girls spent on the investigation into the value of a might have been used to gain other insights. One could also argue that this kind of process, where students encounter a problem and are able to overcome it themselves through investigation and reflection on their process is not a waste of time, but rather is good as students boost their self-efficacy through achieving a solution on their own. The problem of the identity element being omitted, however, might lead to misconceptions that $x = 1$ as described in the literature (e.g. by MACGREGOR & STACEY 1997, see subsection 2.3.3), so it is important that this problem is addressed or overcome by the students in order to avoid the building of these misconceptions.

12.3.3 Parameter Stereotypes

Linear functions are often taught in the form $y = m \cdot x + b$, with m standing for the slope and b for the y-intercept. Emphasizing on these letters for the two parameters can lead to parameter stereotyping by the students, so students might directly and wrongly identify b in $f(x) = a(x - b)^2 + c$ as the y-intercept, as the students over-generalize their knowledge gained with linear functions, specifically their knowledge about the designations of parameters in linear functions.

One of the students showing this kind of parameter stereotyping is Charlotte, who worked with Noah using the function plotter approach (see also subsection 12.1.1). The class was taught linear functions in the form $y = m \cdot x + b$ and the students learned that m is the slope and b is the y-intercept of a linear function. After working on the first two tasks of the worksheet, Charlotte and Noah decide to write down key-points for the three influences (see phase 14 in Figure 12.2 and section F.1). During this the following discussion takes place:

158. Charlotte: wait, we had to do c, right? So c is b is the y-intercept, if the y-intercept changes or in that case is bigger, no yes
159. Noah: yes of the positive one cannot say now just positive
160. Charlotte: From positive uh from negative to positive, uh the graph becomes smaller overall in proportion
161. Noah: yes sharper we say sharper not smaller, because it does not get smaller
162. Charlotte: yes it gets sharper and the y-intercept
163. Noah: moves upwards
164. Charlotte: moves upwards
165. Noah: so (*writes down*) uh how can we write this, basically uh yes key point y-intercept

Charlotte identifies in Line 158 c as the y-intercept, which is correct for the examples they graphed in task 2.1, but she refers this to b in linear functions. This stereotyping probably stems from the teaching of linear functions that are in Germany mostly in the form $y = m \cdot x + b$ and over-generalizing this knowledge (see e.g. BILLS 2001 and subsection 2.2.3). They also think that the graph is narrower if one changes c (in line 161) and with this display a common misconception (see e.g. GOLDENBERG 1988 and subsection 2.2.3) that is based on the wrong interpretation of the given graphical information.

166. Charlotte: the y, when the y, when the y-intercept uh moves upwards, the term exactly the term is sharper, even though the m-value, m-value so the slope does not change, because the slope does not change.
167. Noah: if the y-intercept, how can we say this?
168. Charlotte: I know, I don't know if that's correct, but if the y-intercept changes, in this case then
169. Noah: ah so changes (*still writing*)
170. Charlotte: changes
171. Noah: and then the term uh the graph is sharper
172. Charlotte: yes, even though the m-value does not change, because the m-value does not change and m is technically the slope, that is why I find it strange, but I don't know if that is what we had to find out

Even though Charlotte realises correctly that the value for a does not change (in line 166), in this phase of the intervention the misconception is not treated, and the output of the technology is accepted and the intuiton ignored (as also described by CAVANAGH & MITCHELMORE 2000a, see section 3.2). While she expresses this confusion (in line 166 and again in line 172), it becomes clear that she identifies a with the m in the linear functions, again over-generalising from the linear functions. They only overcome the misconception of the graph changing the slope, when c is changed, later in the intervention, when Noah uses the technology to explain why this assumption is wrong (see subsection 12.2.3).

In the course of the Guided Discovery Charlotte shows the fixed meaning for b as the y-intercept again, while working on the explanation task regarding b. They try to comprehend the two on the worksheet given statements for b (namely *If I look at $(x-1)^2$ instead of x^2 all values in the table move towards bigger x* and *$x^2 = 9$ is correct for $x = 3$, but $(x-1)^2 = 9$ first applies for $x = 4$.*) and recap their insight into the influence of b (see phase 21 in Figure 12.2 and section F.1):

279. Noah: ok wait, give reasons for the influence of b. Yes basically b, yes if you, so the b-value is basically where the term is on the x-axis or not?
280. Charlotte: yes no, on the y-axis, isn't it?
281. Noah: this is the x-axis (*moves pen along the x-axis*), here that is the
282. Charlotte: yes, but b is actually the y-intercept, but they do not mean that, they mean the variable b and not b in that sense
283. Noah: yes I know
284. Charlotte: they mean this here (*taps onto Part 2 on the worksheet*).

Again, Charlotte thinks b has to be the y-intercept as this is how she has learned it and then states in line 282 that now with b something different is meant. It can be conjectured that Charlotte sees m and b in linear functions more like physical unit, so e.g equivalent to A for ampere as the unit for electric current, and not as the mathematical parameter, which could be replaced with any other letter. Even though Charlotte has this strong parameter stereotype, she still manages to correctly discuss the influences in the frame of the given quadratic functions, but it is clear that she has not yet fully understood the variable and parameter concepts, as she does not realise that the assignment of letters can be arbitrary and if two parameters are represented through the same letter, they have not necessarily the same influence. This could be seen as a similar misconception to the *different variables—different values* misconception, where students think that two variables cannot have the same value (described e.g. by WARREN 1999, see also subsection 2.3.3). The stereotyping shown by Charlotte might be a reversal of this misconception in the sense that she thinks the same naming of variable must have the same purpose in different concepts.

The example of Charlotte shows therefore the importance to use different parameters at some point in the learning, so the stereotyping does not occur to this degree. Stereotyping can be useful in the sense of automating procedures and therefore can be helpful at the start of the conceptualization (see e.g. BILLS 1997), but for a flexible understanding of the parameter concept, the students also need to realise that the assignment of a letter for parameters is arbitrary. Settings like the one used in this Guided Discovery enable identification of the parameter stereotypes students might have and thus offer potential for then overcoming these misconceptions and resulting in learning about the parameter concepts.

12.3.4 No Need to Explain

Explaining the influences of the parameters and not just describing insights only occurs in very few processes, for example in the process of Grace and Laura (see subsection 12.2.4). Grace and Laura use the Grundvorstellung mapping to explain the influence through a both using the graph and the function table. But this is an example of very few and most students seem to not explain the influences, which is also obvious in the summary sheets results (see subsection 11.1.7). Most of the students only wrote statements about the influence on their summary sheet and no explanations thereof.

The videos show that students do work on task three, which was designed to evoke explanations, but most students only describe their insights again and the worksheet does not succeed in fostering explanations of the results. This problem can be found in many processes, so for example in the process of Charlotte and Noah (see Phase 20 in Figure 12.2 and section F.1). While they are working on task 3.2 (Explanation for a), Noah thinks they have already done the task.

253. Noah: (*reads out loud*) now try to explain the influence of a. Wait what was a again? Ah there (*reads out aloud what he has written down*) the graph moves on the x-axis but stays the same. No wait what was a?
254. Charlotte: a was b
255. Noah: a was b ok
256. Charlotte: the influence of a is
257. Noah: (*still reading aloud*) look at your findings in exercise 2 again and compare to the standard parabola. Specify exactly how and why the graph changes. Yes we did that already now

Checking with Charlotte and their notes from part two of the worksheet (in line 253) what the parameter a was, Noah states, after reading the task 3.2, that they have already done so (line 257). They then recap their results gained in task 2.2, which is part of the task, but omit explanations.

258. Charlotte: look, look at for this (*incomprehensive*)
259. Noah: yes like if it is a positive value then it goes upwards
260. Charlotte: yes and if it is a negative value
261. Noah: then it is in the positive, then it goes upwards, let's just say, and when, because look here it is as well, this side is negative
262. Charlotte: yes
263. Noah: then it goes upwards and if it is a negative value it goes downwards
264. Charlotte: actually logical and like if it is zero it is zero. Then nothing is, it is not positive or negative. Yes ok then (*reads out*) Finally try to explain the influence of *b*
265. Noah: wait a second (*writes down*)

The insights they recap are viable but rather coarse, as they do not specify when the graph is shrunk or stretched. They only include the concavity of the parabola including the special case $a = 0$, where only a straight line is graphed (in line 264). Even though they explicitly state the influence depending on the sign of a and differ between positive and negative values, there is no explicit explaining of the shrinking or stretching. It might be that they have some kind of explanation in their heads, but they do not state them in the recap, which is all that can be observed. They then proceed to task 3.3.

So either due to the wording of the task or because the students do not know how, they do not succeed in explaining the influence for example through arguing with the multiplication or the Grundvorstellung mapping as Grace and Laura did in their process (see subsection 12.2.4). All videos show however, that students did reach part three of the tasks, so the shortness of the intervention does not seem to be the problem. It seems more likely that the guidance that was implemented in the explanation tasks was not purposeful enough to induce the need to find explanations with all students. The *size of discoveries* design principle (see MOSSTON 1972 and section 4.2) might have been too great for the explanations to be achievable and therefore the tasks do not foster explanations enough.

12.4   Conclusion Video Results

The video analysis gives deeper insight into the processes of students during the Guided Discovery and also shows the variety of how the Guided Discovery processes can progress. The investigation depends on many factors, for example the skills of the students working together, the technological obstacles they encounter and so forth. As students are free to choose the examples they contemplate in their investigation as well as the length of their investigation into each parameter, the two examples in section 12.1 illustrate some of these various views. The Guided Discovery processes are rather dense with lot of different activities and the students work on the discovery for around 50 to 60 minutes on their own and gain viable insights into the parameter influence and even overcome some misconceptions on their own. The analysis also shows that the processes do not proceed in a linear way, but rather the students return to points of their investigation a number of time during the process (e.g. subsection 12.3.2).

Taking the research questions posed in Chapter 5 into account, the video analysis can provide insight into a number of those.

Overall the video analysis shows that the Guided Discovery into the parameter influence is successful to some degree with all students. All students achieve some degree of insights into the influences, so the design principles for Guided Discovery (see MOSSTON 1972, section 4.2) that were implemented to foster this insight seem to be appropriate for the students to master the challenge of discovery.

Regarding the technological subquestion of the research questions (see Chapter 5), a number of affordances and constraints of the used technology can be identified through common aspects in a number of videos. These common aspects show that technology can both facilitate and hinder the Guided Discovery in different ways. The constraints regarding technology identified in this study (subsection 12.3.1) might be avoided if students were more familiar with the software used, while the affordances are even more surprising, when the short time the students worked with the software before the Guided Discovery is taken into account. The three different technological approaches all show both affordances and constraints, but due to the individuality of the students' processes the video analysis does not provide any insight into which approach is the most beneficial in the conceptualization of parameters. The affordances show, however, that students in all groups are able to use the technology as an instrument in the sense of instrumental genesis (as described in section 3.1), for example while testing hypotheses or exploring on their own (subsection 12.2.1).

Using the framework of BALL & BARZEL (2018) the influence of technology on the communication was shown, as the framework enables identification of the influence of technology on the communication. Both communication of technology displays, where the displays are used as a stimuli for discussion and communication with technology through entering syntax (BALL & BARZEL 2018) were expected and are common in the intervention. It was also shown that students even used the technology to explain or show something to the other students including the technology in their communication (subsection 12.2.3).

Regarding the content-related subquestions (see Chapter 5), the video analysis corroborates the summary sheets results that students achieve some degree of conceputalisation of parameters through the intervention. Identifying the influences of

the three parameters is successful and some students even use the technology provided to overcome misconceptions through the possibility to investigate different
examples and change the window of the shown graphs. The implemented guidance
succeeds in guiding students to identify the influences.

Two important aspects of problems regarding mathematical understanding were
also identified in the videos. The problem with the value of a (subsection 12.3.2) in
the standard parabola shows an incomplete understanding of the parameter concept
and can be cleared up by explaining the convention that if a variable is multiplied with
one, the one can be omitted, when writing down. The parameter stereotyping visible
for example in the process of Noah and Charlotte (subsection 12.3.3) probably stems
from teaching linear functions always in the same version $y = m \cdot x + b$, this should be
reconsidered so students develop a flexible and deep understanding of the parameter
concept.

The video analysis also showed that students do not write all their results onto
their final summary sheet, so it can be conjectured that students might have gained
even more insight than analysed in Chapter 11. The summary sheets results showed
that very few students wrote explanations for the influences on their summary sheets
(see subsection 11.1.7), this was also evident in the videos. Even while working on
the tasks that were supposed to foster explanations, many students only described
the influence again and did not explain them (subsection 12.3.4), but some students
were able or at least attempted to explain the influence using the Grundvorstellung
mapping and function tables or even using the graphs (subsection 12.2.4). So in
the case of fostering explanations the implemented guidance (see Figure 7.18 and
subsection 7.2.3) was not effective enough.

Part V
Conclusion

Conclusion 13

The aim of the study was to investigate connections between the conceptualization of parameters of quadratic functions, the use of technology and Guided Discovery. In this conclusion the key results of the presented work will be summarized, the limitations presented and the implications for teaching and learning discussed.

13.1 Summary

There already have been a number of studies investigating discovery learning (see for example ALFIERI et al. 2011), functions (see for example LEINHARDT et al. 1990, KLINGER 2018) or the use of technology (e.g. BARZEL 2006, BARZEL 2012, DRIJVERS et al. 2016) over the years. But the interdependency of the use of technology, Guided Discovery and conceptualization is a complex issue with a lot of affordances and constraints possible. This study aims to give more insight into this interdependency through analysis of processes as well as students' writing products. For this, an intervention study was conducted in which students were guided through the discovery of the parameter influence on quadratic functions of the form $f(x) = a \cdot (x - b)^2 + c$ using a number of tasks. Different versions of technological visualization were chosen in order to investigate the influence of the technology. Apart from the different kind of visualization, the intervention was the same for all four groups and structured in four parts with an introduction that stated the aim of designing a summary sheet at the end of the intervention:

1. First students were asked to describe the differences between the graphs of the functions $f(x) = x^2$ and $f(x) = 4 \cdot (x - 2)^2 + 3$.

L. Göbel, *Technology-Assisted Guided Discovery to Support Learning*, Essener Beiträge zur Mathematikdidaktik, https://doi.org/10.1007/978-3-658-32637-1_13

2. Then the students investigated the graphs if a parameter is changed. For this, they were provided with a possible structure of $f(x) = x^2 + c$, $f(x) = a \cdot x^2$ and $f(x) = (x - b)^2$, they could use to organize their work.
3. Fostering explanations was the intended goal for the third part of the intervention task, where the students were provided with fictitious and conflicting statements regarding the influence of c, the hint to compare their chosen examples for a with the standard parabola and two correct explanations for parameter b and asking for further explanations.
4. The last part of the intervention was intended to reorganize the students' findings and to foster reflection. For this, students were asked to write summary sheets in pairs, which were then collected.

The kinds of technological visualization that were compared in this study were a control group without the possibility to graph functions, a standard function plotter without any programming, and two versions of preprogrammed files in a multiple-representation system (MRS). The two versions of the MRS differed in the way students were to manipulate graphs. While the drag mode allowed students to manipulate a function directly by dragging the graph, in the last kind students were to manipulate the parameters through sliders.

A total of 14 classes with 383 students in total took part in the study. In order to generate a baseline for comparison a pretest was implemented before the intervention. The pretest results (Chapter 10) confirm that the four intervention groups can be compared, as their mathematical achievements do not differ significantly. As a data basis for the analysis the summary sheets of 353 students were collected and two students each in 13 of the 14 classes videographed.

The key results of this work can be summarized and discussed using the answers to the research questions posed in Chapter 5:

Technology-related subquestions

What influence of the different technology use can be identified on the summary sheets?

The analysis of the summary sheets the students produced in the intervention led to the identification of differences between the different groups (see Chapter 11). Since the tasks the students worked on, were the same in all four groups it can be reasoned that these differences stem from the kind of visualization used, while working on the tasks.

If one only differs between the two dynamic (*drag mode* and *sliders* group) and the two static groups (*function plotter* and *without visualization* group), it seems that the dynamic groups foster the learning more than the static. This is evident in the statistical analysis if split only between dynamic and static groups (see Appendix E for exact results). Also there is a tendency that dynamic groups use more dynamic language than the static groups (section 11.2).

This view is too coarse however, as the detailed results (see section 11.1) differ not only depending on the groups, but also which parameter and which influence of the parameter one looks at. A comprehensive overview of the results split between parameters and intervention groups can be found in Table 11.14.

The summary sheet analysis can lead to the conclusion that the drag mode is best suited for the investigation of the vertical stretch and shrink and the vertical transformation, while sliders are best suited for the investigation of the reflection through a and the horizontal transformation (see Chapter 11).

The advantage of the *drag mode* group for the vertical transformation and vertical shrink and stretch might be explained due to a lesser cognitive load. The students in the *drag mode* group manipulate the function directly at the graph and a corresponding equation is shown. On the other hand students in the *sliders* group manipulate sliders for each parameter which then in turn change the graph. The slider could also be seen as an additional step in the change of representation as the numerical value of the slider is not substituted in the general equation displayed. So students in the *sliders* group need to realise and understand which change in which slider results in what change of the graph. This might provide an obstacle in the case of c, when one moves the slider left and right, but the graph moves vertically. The slightly lower number of viable statements by students in the *sliders* group regarding this vertical transformation could stem from this obscuring and would substantiate the conjecture by ZBIEK et al. (2007). This connection might be easier in the *drag mode* group, where they directly see the result of their dragging in the function equation.

But it could also be argued that the slider acts as a direct external representation of the parameter as a changing quantity (as described for example by DRIJVERS 2003) and enable the students to connect the change in the parameter with the change of the graph. Especially in the case of horizontal transformations this might prove to be an advantage over dragging directly at the graph and therefore explain the differences in the results. Also when the graph is dragged horizontally, in the function equation displayed in the *drag mode* group file the equation is simplified for negative values of b, so $(x - (-1))^2$ is simplified to $(x + 1)^2$. It is then the problem of direction in the horizontal transformation described in the literature for example in ZAZKIS et al. (2003). With the sliders this confusion should not appear, so if students move the slider for b to a negative value, the parabola moves to the left of the y-axis.

Students in the *function plotter* group had to actively choose the examples to plot and the results surprisingly identify this approach as the least suitable compared to the other three. In case of the vertical stretch, vertical shrink and reflection through *a*, even less viable answers are on the summary sheet than in the *without visualization* group. It can be conjectured that students in the *without visualization* group chose to sketch the extreme values that were given on the worksheet first and then identifying the influences, whereas students in the *function plotter* group might have plotted not important enough examples to gain insight.

The influence of the technology is also visible in the technical terms used (subsection 11.1.6). 39.4% of students in the *without visualization* groups (which only used scientific calculators) used the term *slope* on their summary sheets in a nonviable way, while only 11.8% of students in the *drag mode* group used the term slope at all on their summary sheets. The terms *x*- and *y*-axis are used more in the experimental groups, especially in the *drag mode* group. The difference regarding *x*- and *y*-axis probably stems from the intervention design, as the three experimental groups used technological visualizations and thus the students have to interpret the results verbally.

What affordances and constraints of the technology are visible?

The results above show that there are differences between the four intervention groups, which can be attributed to the technology. But there are also common aspects regarding the three technological kinds of visualization in this Guided Discovery setting that were identified through the video analysis. These can be clustered into affordances and constraints of the technology.

During the intervention the students in the three experimental (so *function plotter*, *drag mode*, and *sliders* group) were provided with a multiple representation system in different configurations. These can be seen as slightly different didactic configurations.[1] In the classroom these didactic configurations might then be used for different purposes and thus result in different exploitation modes (as described by DRIJVERS et al. 2010). As the students work on their own, it can be argued that they take over the role of their teacher and choose the exploitation modes themselves.

Three main affordances through the use of technology were identified in this study. These are on different levels, while the use of technology as a visual aid is more on an organisational level of the work process, the other two uses of technology

---

[1]In the sense of instrumental genesis as described in section 3.1 and by e.g. DRIJVERS et al. (2010).

(testing of hypotheses and exploration of own examples) are directly related to the gaining of the insight (subsection 12.2.1).

Using the technology as a visual memory aid to redisplay the functions that were worked on earlier helps the students in their non-linear process of the discovery, as the students can revisit the examples they previously worked on. This memory aid might also provide some non-directive support (as described by DE JONG & NJOO 1992) as it helps to organize their process. So in this case the students choose to use the technology as an organisational help.

When students use the technology to test their hypotheses, it matches the way technology is used by experimental mathematicians to prove theorems, when the mathematicians use the technology "to test and especially falsify conjectures" (SINCLAIR 2004, p. 235). The use of technology to explore on their own that was identified in this study can also be linked to the experimental mathematicians' process. This use can be seen as a combination of the three uses "to gain insight and intuition", "to produce graphical displays that can suggest underlying mathematical patterns", and "to discover new patterns and relationships" (SINCLAIR 2004, p. 235). So students with no great experience in discovering mathematical ideas through technology use proceed in the same way as mathematicians might observe, while trying to prove a theorem.

In all three affordances described in this study, students use different aspects of the multiple representation system with an intended aim. So students use the technology as an instrument (as described e.g. by DRIJVERS & TROUCHE 2008, see section 3.1). Different kinds of uses are visible and they might be seen as different exploitation modes, especially the difference between using the technology for organisational aspects compared to using the technology to gain insight. This all occurs, even though most students only worked with the used software for the second time, so it is a promising result. This could be due to the familiarity with digital technologies in their daily life (FEIERABEND et al. 2016, FEIERABEND et al. 2018), as students are used to working with technology, i.e. tablet computers or smartphones. So the students were already familiar with the use of technology in general and only had to acquaint themselves with the software used. This can lower the threshold for the beneficial use.

However, some obstacles were also evident as students do not know how to operate the software or are confused through some aspect of the technological display (subsection 12.3.1). The first obstacle probably stems from unfamiliarity with the software and might be prevented if students use it more and thus develop more usage and instrumented action schemes (TROUCHE 2005a). This problem might be an explanation for the lower viable statements of the *function plotter* group as it is the least intuitive of the three experimental groups. Students in the *drag mode* and *sliders* group only had to drag at some aspect of the MRS, which is more intuitive than explicitly entering function equations and plotting them. So the problem of not knowing how to operate the software most likely occurs more often in the *function plotter* group.

The second problem, which can occur in any of the three experimental groups, can serve as a starting point for discussion, but might need guidance and explanations through the teacher to overcome this problem of unfamiliar technological displays. The technological displays can also evoke misconceptions, e.g. in the case of parameter c the misconception that the graph is narrower for greater values of c (described e.g. by GOLDENBERG 1988). But some students who show this misconception during the investigation succeed to overcome it with the help of technology through revisiting it again over the course of their non-linear process (see for example Charlotte and Noah, subsection 12.2.3).

In total, the affordances that were observed during the intervention might explain the rather good results achieved especially by the *drag mode* and *sliders* group on the summary sheets regarding the conceptualization of the parameters.

Content-related subquestions

What conceptualization of parameters takes place?

During the intervention the students were asked to investigate the influence of the parameters a, b, and c in $f(x) = a \cdot (x - b)^2 + c$ and explain these influences. As these influences can be transferred to higher-order polynomials, they serve as a blueprint for parameter influence in general. Thus, quadratic functions can be used as a starting point for the conceptualization of parameters. So it can be argued that if students achieve in gaining insight into parameters of quadratic functions, they also gain insight into a further, important but even wider concept namely parameters in general.

The summary sheet results show that in all four groups in the intervention, so the control group and the three experimental groups, a large number of students stated viable insights for all three parameters, so the Guided Discovery approach is successful in fostering the investigation (see Chapter 11). The results show that 84.7% of the students stated viable aspects regarding the vertical transformation through parameter c and 77.1% stated viable answers regarding the horizontal transformation through parameter b. This concurs with the literature in the sense, that the vertical transformation is perceived to be easier than the horizontal one (e.g. EISENBERG & DREYFUS 1994, BAKER et al. 2000, GADOWSKY 2001).

The vertical stretch and shrink through parameter a in this study seems to be the most difficult as the overall percentages of viable answers is lower in all analysed categories (subsection 11.1.3) than the viable answers in the vertical and horizontal transformation categories through c and b (subsection 11.1.4 and 11.1.5). It substantiates the findings of KIMANI (2008) that scaling transformations are more difficult for students than translations. In the case of parameter b and c the students could observe the functions with the Grundvorstellung object as a whole (as described by VOM HOFE & BLUM 2016) and the complete graph is moved, while in the case of a the shape of the graph changes. To specifically describe the influence of a one might need to observe the function using the Grundvorstellung covariation.

Regarding parameter a, the video analysis revealed a problem, which also might explain the lower number of viable statements. Due to the wording of the tasks in part two of the intervention, the values of the three parameters were not written, if they could be omitted. For example the function $f(x) = x^2 + c$ was written for part 2.1 instead of $f(x) = 1 \cdot (x - 0)^2 + c$. Also, the standard parabola was introduced and referred to as $f(x) = x^2$. This leads to the problem that students did not remember, what the value of a in the standard parabola is and this can lead to problems while investigating the influence of a (see subsection 12.3.2). In order to avoid this problem, the value of a in the standard parabola should be addressed and explained to the students. This might avoid the obstacle the students encountered as well as simplify the comparison for the explanations. Some students, like Rose and Nina, who encounter this problem are even able to overcome this problem on their own through revisiting it a number of times and resolving it through investigation using the technology (see subsection 12.3.2).

During the conceptualization the students encountered a number of obstacles and misconceptions, for example through over-generalizing from linear functions, which has also been described in literature (see e.g. ZASLAVSKY 1997, BILLS 1997, ELLIS & GRINSTEAD 2008). Probably due to teaching with stereotyping, students sometimes fixate on the allocation that b must be the y-intercept (see subsection 12.3.3), and this can hinder the flexible insight into the parameter concept. Some students also use the term *slope* on their summary sheet and most of these do so in a non-viable way (see subsection 11.1.6) as they over-generalize from linear functions.

Identifying the influences correctly is only a first step in the conceptualization, in order to substantiate the knowledge explaining the influences should be aimed for. But the study here has shown that evoking explanations can be seen as a special challenge. The third part of the intervention was intended to address this challenge and evoke explanations and a reflection of the students' insights. Across all four groups, explanations and reasons were hard to find and very few students succeeded in trying to find correct reasons (subsection 11.1.7). The reasons and explanations given often focussed on the Grundvorstellung mapping (see subsection 12.2.4). Taking a closer look at the students processes while working on the explanation task, it can be seen that students might not even see a necessity to explain their insights while working on the task, so the small number of explanations might be due to the posing of the task (see subsection 12.3.4) or that students are not used to explaining influences in class. It can be concluded that for evoking explanations even more guidance than already implemented is necessary and the Guided Discovery approach in this study might not be suited for evoking explanations. For this, an approach with more teacher involvement might be needed.

Some degree of conceptualization took place in all groups regardless of the technology used. But the technology might also have an influence on the students' learning, which is the focus of the next research question:

How does the technology use influence the students' learning?

Students in the three experimental groups, so *function plotter, drag mode* and *sliders* group used the technology in their investigation into the parameter influence. The affordances and constraints described above have already given some degree of insight into how the technology influences the learning. The two examples of Guided Discovery processes (section 12.1) can show to some degree the variety of how students use the technology in their investigation. The three technological features investigated in this study can potentially foster the discovery in different ways. In all three cases the procedural skills necessary to graph functions per hand are offloaded to the technology, which can lead to increased conceptual understanding (e.g. KIERAN & DRIJVERS 2006) and in all three features the possibility to show multiple representations of a function at the same time is available.

The function plotters in this study offer the possibility to plot more than one function at the same time. This feature can simplify the comparison between the different functions for example in the investigation into the influence of parameter a. Students can directly compare the graphs and the corresponding tables (see for example subsection 12.2.4) and might even use this feature of directly comparing two graphs to overcome a misconception (Phase 25 in Figure 12.2 and section F.1). But students have to decide for themselves which functions to plot and when to display a function table. This openness in their decision can lead to the problem that students do not investigate enough or meaningful examples and therefore offers less pre-structure and guidance than the two dynamic features. This might explain the lower number of viable insights on the summary sheets that students in the *function plotter* group noted.

With the drag mode, students can manipulate a function directly through dragging at the graph. The given file also included a linked function equation and function table that was visible at all times. The three representations are dynamically linked and therefore provide the "hot link" described by KAPUT (1992) which can support understanding. The direct dragging is intuitively managed by the students and in the case of a and c the results of the study lead to the conclusion that the drag mode seems to be the most suitable for the investigation. In the software used in this study, it is hard to manipulate only one parameter at the time or to achieve integer values for the parameters through dragging. The latter might lead to some confusion, when working with the function table as the function values are therefore rather difficult to interpret. The drag mode might also be used to focus the change of representations from graph to term (see subsection 12.2.2), rather than the change from term to graph that is emphasized in the function plotter approach.

The last technological feature chosen for this study were sliders. With this approach, the students were able to manipulate one parameter at the time through sliders. This enables students to distinguish between the different influences and identify the three influences simply through moving the sliders (for example the case of Inga and Tine in subsection 12.2.1). In the case of parameter b, sliders seem to be the most suitable approach to investigate this, as the horizontal movement of the sliders corresponds with the horizontal movement of the graph. The slider might act as a further external representation of the parameter and directly representing parameters as a changing quantity (as described by DRIJVERS 2003) and thus fostering the insights. As with the drag mode, a linked function table was visible in the given file. However, for some students the movement of the slider as a way to manipulate the parameter with the resulting change in the graph might obscure the connection (ZBIEK et al. 2007) or lead to superficial investigation (DRIJVERS 2004). In the case of c the perpendicular movements of slider and parabola might outweigh the benefit of the additional external representation of the parameter. This would need to be investigated for example through programming the slider vertically and testing if this leads to even better results regarding parameter c.

The three approaches have different aspects that might influence the learning, but in all three cases the technology can influence the communication between the students:

Communication-related subquestions

Which kind of communication while using technology occurs during the intervention?

During the intervention the students worked in pairs, so communication necessarily occurs. Using the framework of communication while working with technology (BALL & BARZEL 2018), three kinds of communication have been identified. Students communicate with the technology through input of syntax or manipulating the graphs. Also, the communication of technology (BALL & BARZEL 2018, p. 233) occurs when the display provides a stimulus for discussion through non-intuitive or surprising results. Finally, a combination of the two occurs when students interact with the technology to explain or discuss something with their partner and therefore including the technology in the communication (see subsection 12.2.3).

Peer interaction which can be observed for example through the communication including technology can also achieve to move the target of the discovery into the students Zone of Proximal development (VYGOTSKY 1978) and therefore contribute to the success of the Guided Discovery. So it can be concluded that communication is a key factor in this Guided Discovery and working in pairs was a valid choice to achieve this.

The answers to the five subquestions can be combined to answer the main research question of the study presented in this dissertation:

How does technology-assisted Guided Discovery influence the conceptualization of parameters of quadratic functions?

Technology-assisted Guided Discovery can benefit the conceptualization of parameters in quadratic functions. The approach chosen for this study can guide the students through their investigation and leads to some degree of viable conceptualization. The guidance succeeded in fostering the students to identify the influences into the parameter, so the implemented series of tasks (see subsection 7.2.3) with the help of the technological visualization (see section 6.4) moved this into the students' Zone of Proximal Development. The assistance provided through the technology also attributes to the successful investigation, as it provides a number of affordances and also influences the communication during the intervention. The dynamic visualizations, so the use of drag mode and sliders, seem to be better suited for this investigation than the static one using function plotters or the group using no technological visualization. Achieving to explain the influence of parameters, however, is not as easy and the technology-assisted Guided Discovery implemented in this study does not succeed in fostering this regardless of the technology used. This is one of the limitations of this study, others will be presented in the next section.

13.2 Limitations of the Study

The study presented in this thesis implemented technology-assisted Guided Discovery with the narrow focus on parameters in quadratic functions. There are a number of limitations of this study that have to be taken into account regarding the results itself and their transferability to other aspects of mathematics learning.

The intervention conducted here tried to cover a topic in 90 minutes which usually is taught over a longer period. One could argue that the conceptualization of parameters is not successfully possible in this short time and only using Guided Discovery without input by the teacher. But the study showed that this is not the case. Conceptualization is possible in this short time, at least to a certain degree and students do succeed in investigating this on their own.

The four kinds of visualization compared in this study might differ in its suitability, however the statistical analysis only shows that there are significant differences in the percentages and not which approach is the best. Only through qualitative interpretation of these percentages one can conjecture that the dynamic groups are better. For a definite answer, which approach is the best, a larger sample size and a post-test that measures the knowledge gain would be needed.

Also, explanations are obviously too hard to achieve in the short time of the intervention. This might be due to not enough scaffolding in the tasks or because students have not achieved a sufficient amount of learning gain. It could also be that the students were not used to explaining their insights and did not know what they have to do if the task asks to explain the influence. This problem might stem from a missing culture of argumentations in mathematics classes, so students would only be used to solve mathematical problems and not argue intensively. So in order to evoke the explanations, even more guidance should be implemented in the task regarding the explanations. This is also a starting point for the teacher to intervene and collect the students' results from their investigation and then lead the students to explaining the influences.

Most of the results presented here are based on the analysis of the summary sheets. It can be argued (and this is to some degree visible in the videos) that students do not write all their results of the exploration on the summary sheets. An example is the process of Charlotte and Noah (subsection 12.1.1), where they do not write their insight into $a = 0$ down. This does not weaken the results that the Guided Discovery achieves some conceptualization of the parameters, but rather strengthens them as the students might have found out even more than identified through the summary sheet analysis. In order to identify the actual degree of conceptualization that took place, a more extensive video study as well as a pre-post-test design that ascertains the degree of conceptualization should be conducted.

An important point to consider is the transferability of the results to other mathematical concepts. It can be assumed that for similar investigations like higher-order polynomials the Guided Discovery developed in this thesis can also provide a suitable approach. Regarding the three technological visualizations, the transferability heavily depends on the complexity of the mathematical concept. Sliders offer the most pre-structure for parameter investigation, but with more complex investigations the students might only investigate superficially (as described by DRIJVERS 2004). The drag mode is most likely the least transferable of the three, as it is only available in selected software. Also the drag mode in TI-Nspire does not work with polynomials with a degree greater than two. It does, however, work for example with trigonometric functions. So the transformation of trigonometric functions is an example of a concept, where the approach presented here might be used successfully. The function plotter is the most transferable of the three technological visualizations, as function plotters are widely available and pose no limitation on the kind of function that is graphed, but might offer too much choice in the numerous examples and thus might lead to confusion (BARZEL & GREEFRATH 2015).

13.3 Implications

The aim of the dissertation was to give further insight into the connection between technology, Guided Discovery and conceptualizing parameters in quadratic functions. A number of implications for future research as well as the teaching and learning can be drawn from these results and these will be presented here.

The Guided Discovery approach is successful and the structured worksheet helps the students to gain viable insights regarding the parameter influence. The four parts used to guide the students enabled to open the possibility of discovery for them. This approach could be adapted for other aspects of mathematical understanding and then implemented into teaching. Especially at a start of teaching a new topic, this Guided Discovery might motivate the students and therefore foster the learning. Testing this approach in other fields and comparing these results might also lead to identification of more design principles regarding the Guided Discovery and this could attribute to the discussion of discovery learning and its efficacy in general.

Regarding the Guided Discovery approach, there are still a number of aspects that were not investigated in the study here, but might provide valuable insights into the learning using Guided discovery. Apart from the large-scale testing of conceptualization and therefore the efficacy of this approach described above (section 13.2), the interdependency of the specific task, the technology and the learning process shows a promising field to investigate. For this the theoretical approach of embodied cognition (e.g. RADFORD 2014) might prove a valuable basis for this investigation. Especially while using the drag mode, embodied cognition might play a substantial role. The study also shows that one should carefully consider what aspects of the technology might influence the learning in what way and how this matches the intended learning goal.

The teaching of parameter concept usually takes place over a number of weeks. The study presented here shows that a first discovery of the parameter influence can take place in a very short time and students find out many viable insights. This would obviously have to be systemized in class afterwards, so the non-viable discoveries are corrected, but the first discovery might motivate the students for the consolidation of the concepts. The investigation into parameters using digital tools was implemented in the revised state curriculum in Germany for students starting high school after 2019 (MSB NRW 2019, p. 33), so the approach presented here would directly address this aspect of the curriculum. In order to achieve more explanations during the discovery, the third task of the intervention should be redesigned, so students would be more guided towards explanations. It is important, however, to find a balance between enough guidance, so students achieve the goal and enough openness to enable students to discover and follow their own learning pathways.

Using the technology during the investigation is important and might even convince teachers of the affordances of using technology. But it is also visible in the results that students should work with technology on a regular basis in order to develop the instrumented action schemes necessary for a successful use. For this, there is a need of professional development for the teachers so that the use of technology is implemented adequately. The videos recorded during the intervention could be used for professional development of teachers for case-based learning (see for example EBERS et al. 2019). With case-based learning teachers could be sensitized for the problems with guided discovery and the use of technology in general. This could lead to an even more effective posing of the questions and transfer of the gained knowledge to other topics.

The thesis opened with the quote "(...) there is no understanding without visualization" (DUVAL 1999, p. 13). The study has shown that for the case of parameters, visualization plays a crucial role in the understanding of parameters. The hierarchy between parameters and variables was however not investigated, but might offer promising fields to investigate. The implications illustrate a number of fields where further research could follow this dissertation. There are still a number of questions unanswered and open for newer research. The thesis closes with the statement of some of these questions.

- How does the learning using the different technology progress in detail?
- Does the Guided Discovery provide an approach for every mathematical concept?
- How can explanations be evoked with the students?
- How can teachers be prepared to implement digital tools and discovery learning successfully?

Original German Versions of Pretest Items

The pretest items shown in section 7.1 were translated from the original german items, which are shown in this chapter.

The original German versions of Item R4TG can be found in KLINGER (2018, p. 301) and of Item N1FQ in KLINGER (2018, p. 243).

© The Editor(s) (if applicable) and The Author(s), under exclusive license to Springer   219
Fachmedien Wiesbaden GmbH, part of Springer Nature 2021
L. Göbel, *Technology-Assisted Guided Discovery to Support Learning*,
Essener Beiträge zur Mathematikdidaktik,
https://doi.org/10.1007/978-3-658-32637-1

Vermerk: P1ZC

Im folgenden Graph sind vier Punkte auf den Funktionen eingezeichnet.

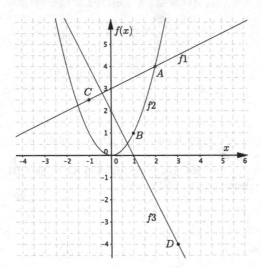

(a) Lies die Koordinaten der Punkte A, B, C und D ab und notiere sie.

A (,   )
B (,   )
C (,   )
D (,   )

(b) An welchen Stellen x nehmen die Funktionen den Wert 1 an?

$f1$:

$f2$:

$f3$:

Figure A.1 German Version Item P1ZC

Vermerk: U2PT

Es gilt $a + b = 35$. Was ist dann $2 \cdot a + 2 \cdot b + 4$?

Antwort:

Figure A.2 German Version Item U2PT, translated from FOY et al. 2013, p. 57

x und y sind beliebige Zahlen. Es gilt $y = 5 - 2 \cdot x$. Was passiert mit y, wenn x größer wird?

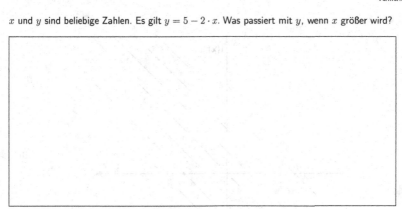

Figure A.3 German Version Item D3JG

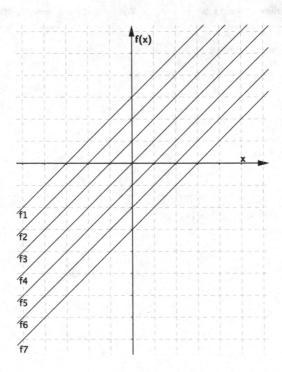

Durch welche Funktionsgleichung können alle Graphen beschrieben werden?

☐ $f(x) = a \cdot x + 2$ ☐ $f(x) = x + b$ ☐ $f(x) = a \cdot x$ ☐ $f(x) = 2 \cdot x + 3$

Begründe, warum nur deine Wahl die Graphen beschreiben kann.

Figure A.4 German Version Item Y3GJ

Vermerk: E4FV

(a) Bestimme die fehlende Zahl a.

$6 \cdot a + 3 = 45$

Antwort:

(b) Erkläre ausführlich, wie du vorgegangen bist.

Figure A.5 German Version Item E4FV

Vermerk: R5TG

Gegeben ist die Funktion $f(x) = a \cdot x + b$, wobei a und b reelle Zahlen sind. Entscheide für jede Aussage, ob sie wahr oder falsch ist.

	wahr	falsch
Für $a = 0$ ist der Graph parallel zur y-Achse.	□	□
b gibt den y-Achsenabschnitt an.	□	□
Für $a = 1$ und $b = 0$ entsteht die 1. Winkelhalbierende.	□	□
Die Funktion $g(x) = -a \cdot x + b$ sieht aus wie f, nur an der x-Achse gespiegelt.	□	□
Für x muss man bei der Funktion alle Zahlen einsetzen.	□	□
a und b stehen für Zahlen, die bei einer bestimmten Funktion fest gegeben sind.	□	□

Figure A.6 German Version Item R5TG

Unten siehst Du die Funktionsgraphen, Wertetabellen und Funktionsterme von drei Funktionen.

(a) Ordne zu welcher Graph zu welcher Tabelle und zu welchem Term passt. Verbinde diese mit einer Linie.

(b) Fülle zusätzlich die leeren Felder in der Wertetabelle aus.

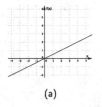
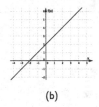
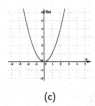

(a)　　　　　　　　　　(b)　　　　　　　　　　(c)

(1)	
x	f(x)
-6	
-5	25
-4	16
-3	9
-2	4
-1	1
0	0
1	1
2	4
3	9
4	16
5	25
6	

(2)	
x	f(x)
-6	
-5	-3
-4	-2
-3	-1
-2	0
-1	1
0	2
1	3
2	4
3	5
4	6
5	7
6	

(3)	
x	f(x)
-6	
-5	-2.5
-4	-2
-3	-1.5
-2	-1
-1	-0.5
0	0
1	0.5
2	1
3	1.5
4	2
5	2.5
6	

(i)
$$f(x) = \tfrac{1}{2} \cdot x$$

(ii)
$$f(x) = x^2$$

(iii)
$$f(x) = x + 2$$

Figure A.7 German Version Item A6XY

Full Tasks

B

The intervention conducted in this study consisted of three 45 minutes lesson. While in the first lesson the TI-Nspire CX CAS was introduced to the three experimental groups, the second and third lesson a Guided Discovery was conducted. The design of the three versions of Guided Discovery (minimally, intermediate and final guided version) was described in section 7.2. The German original tasks of these three versions can be found in subsection B.2.2 for the minimally guided, subsection B.2.3 for the intermediate and subsection B.2.4.

The tasks for the introduction into TI-Nspire were not described in detail above, but they can be found in the next section B.1.

B.1   Tasks of Introduction TI-Nspire

The first lesson of the intervention was implemented to introduce the TI-Nspire CX CAS to the students, who have not worked with it previously (see section 6.3). The lesson was specific for the experimental groups, so students in the *function plotter* groups plotted functions at the same time, students in the *drag mode* group manipulated the graphs through dragging and students in the *sliders* group used sliders. The translated tasks for the respective groups can be found in subsections B.1.1, B.1.2, B.1.3, while the german originals can be found in B.2.1.

© The Editor(s) (if applicable) and The Author(s), under exclusive license to Springer   225
Fachmedien Wiesbaden GmbH, part of Springer Nature 2021
L. Göbel, *Technology-Assisted Guided Discovery to Support Learning*,
Essener Beiträge zur Mathematikdidaktik,
https://doi.org/10.1007/978-3-658-32637-1

B.1.1   Function Plotter Group

Introduction into TI-Nspire

Task 1: Open 2: Eigene Dateien and then EinführungTI and adjust the window settings, so that your function looks like the one in the screenshot.

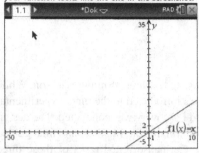

- Instead of a comma you have to write a dot in decimal numbers.

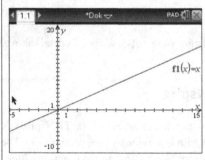
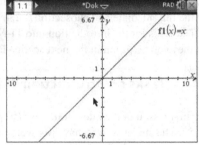

Task 2: Display the entry line with ⊕. With the arrows you can scroll through more functions. Plot the following functions, so that they are displayed at the same time.

- $f1(x) = x + 1$
- $f2(x) = -x$
- $f3(x) = 2x$

Task 3: Delete f2(x) and f3(x) and create the function table for f1(x)=x+1 with ⊕, 7: table 1: split-screen table.

Figure B.1 Translated Task for the Introduction TI-Nspire *Function Plotter* Group

B.1.2   Drag Mode Group

Introduction into TI-Nspire

Task 1: Open 2: Eigene Dateien and then EinführungTI and adjust the window settings, so that your function looks like the one in the screenshot.

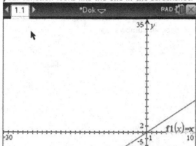

- Instead of a comma you have to write a dot in decimal numbers

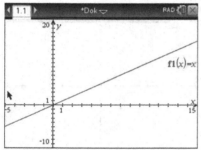

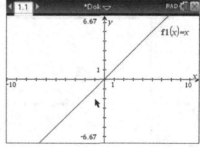

Task 2: Move the cursor onto the function. You can grab the function with ⌨⌘ and drag it. If you grab close to the y-intercept, the y-intercept changes, if you drag somewhere else, the slope changes. Change the function with the touchpad to the following functions.

- $f1(x) = x + 1$
- $f1(x) = -x$
- $f1(x) = 2x$

Task 3: For the function f1(x)=2x create a function table with the menu key, then table, split-screen table.

Figure B.2 Translated Task for the Introduction TI-Nspire *Drag Mode* Group

B.1.3   Sliders Group

Introduction into TI-Nspire

Task 1: Open 2: Eigene Dateien and then EinführungTI4 and adjust the window settings, so that your function looks like the one in the screenshot

- Instead of a comma you have to write a dot in decimal numbers.

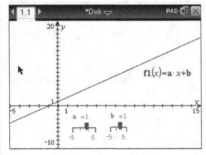
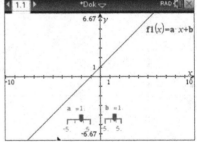

Task 2: Click with the touchpad on one of the slider bars. Now you can manipulate the value of a or b with ◀ and ▶. Change the function with the slider bars, so that following functions are graphed.

- $f1(x) = x + 1.5$
- $f1(x) = -x$
- $f1(x) = 2x$

Task 3: For the function f1(x)=2x create a function table with ⬡, 7: table, 1: split-screen table.

Figure B.3 Translated Task for the Introduction TI-Nspire *Sliders* Group

B.2   Original German Versions of Intervention Tasks

B.2.1   German Tasks Introduction TI-Nspire

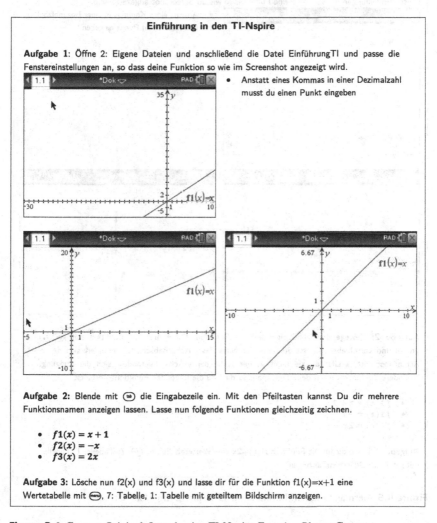

Einführung in den TI-Nspire

Aufgabe 1: Öffne 2: Eigene Dateien und anschließend die Datei EinführungTI und passe die Fenstereinstellungen an, so dass deine Funktion so wie im Screenshot angezeigt wird.

- Anstatt eines Kommas in einer Dezimalzahl musst du einen Punkt eingeben

Aufgabe 2: Blende mit ⊕ die Eingabezeile ein. Mit den Pfeiltasten kannst Du dir mehrere Funktionsnamen anzeigen lassen. Lasse nun folgende Funktionen gleichzeitig zeichnen.

- $f1(x) = x + 1$
- $f2(x) = -x$
- $f3(x) = 2x$

Aufgabe 3: Lösche nun f2(x) und f3(x) und lasse dir für die Funktion f1(x)=x+1 eine Wertetabelle mit ⊕, 7: Tabelle, 1: Tabelle mit geteiltem Bildschirm anzeigen.

Figure B.4 German Original: Introduction TI-Nspire *Function Plotter* Group

Einführung in den TI-Nspire

Aufgabe 1: Öffne 2: Eigene Dateien und anschließend die Datei EinführungTI und passe die Fenstereinstellungen an, so dass deine Funktion so wie im Screenshot angezeigt wird.

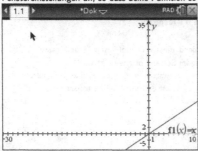
- Anstatt eines Kommas in einer Dezimalzahl musst du einen Punkt eingeben

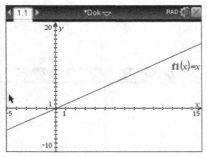 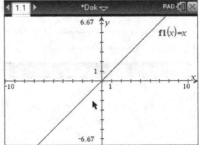

Aufgabe 2: Bewege den Mauszeiger auf die Funktion. Du kannst die Funktion mit ⓒⓣⓡⓛ🔲 greifen und verschieben. Greifst du sie in der Nähe des y-Achsenabschnitts verändert sich der y-Achsenabschnitt, greifst du sie nicht in der Nähe der y-Achse, verändert sich die Steigung. Verändere die Funktion mit dem Touchpad, so dass du die folgenden Funktionen erhältst.

- $f1(x) = x + 1$
- $f1(x) = -x$
- $f1(x) = 2x$

Aufgabe 3: Lasse dir für die Funktion f1(x)=2x eine Wertetabelle mit 🔘, 7: Tabelle, 1: Tabelle mit geteiltem Bildschirm anzeigen.

Figure B.5 German Original: Introduction TI-Nspire *Drag Mode* Group

Einführung in den TI-Nspire

Aufgabe 1: Öffne 2: Eigene Dateien und anschließend die Datei EinführungTI4 und passe die
Fenstereinstellungen an, so dass deine Funktion so wie im Screenshot angezeigt wird.

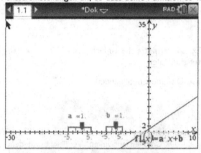

- Anstatt eines Kommas in einer Dezimalzahl
  musst du einen Punkt eingeben

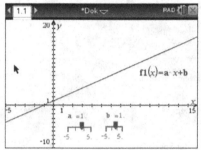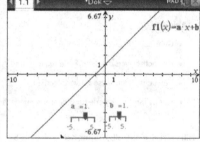

Aufgabe 2: Klicke mit dem Mauszeiger auf die Schieberegler. Nun kannst Du den Wert mit ◀ und
▶ den Wert für a oder b einstellen. Verändere die Funktion mit den Schiebereglern, so dass du die
folgenden Funktionen erhältst.

- $f1(x) = x + 1.5$
- $f1(x) = -x$
- $f1(x) = 2x$

Aufgabe 3: Lasse dir für die Funktion f1(x)=2x eine Wertetabelle mit ⊞, 7: Tabelle, 1: Tabelle
mit geteiltem Bildschirm anzeigen.

Figure B.6 German Original: Introduction TI-Nspire *Sliders* Group

B.2.2   German Tasks Minimally Guided Approach

Bitte beachte bei der Bearbeitung der Aufgaben folgendes:

- Nutze für Notizen die bereitgestellten Blätter.

- Schreibe bitte möglichst leserlich.

- Nutze bitte **keinen** Tintenkiller oder Tipp-Ex, sondern streiche einfach durch. Wir möchten wissen, welche Überlegungen bei der Bearbeitung ablaufen.

- Es wäre hilfreich, wenn du sagst, was du denkst. Diskutiert miteinander!

Aufgabe

Seien $a, b, c \in \mathbb{R}$. Untersuche, wie sich die Veränderungen der Parameter a, b und c der Funktion $f(x) = a \cdot (x - b)^2 + c$, $x \in \mathbb{R}$ auf den Verlauf des Graphen auswirken. Vergleiche mit dem Verlauf des Graphen der Normalparabel $g(x) = x^2$.

Anleitung:

- Geht schrittweise vor.

- Betrachtet verschiedene Fälle für a, b und c.

- Versucht dann, diese zu verallgemeinern.

- Könnt ihr eure Überlegung auch rechnerisch beweisen?

Hilfsmittel Kontrollgruppe: Keine. Erstellt Skizzen für Spezialfälle.
Hilfsmittel Funktionsplotter: Funktionenplotter „Grapher".
Ihr könnt mehrere Funktionen gleichzeitig zeichnen lassen, in dem ihr im Grapher unten links auf das Pluszeichen und neue Gleichung auswählen klickt.
Hilfsmittel Zugmodus: TI-Nspire CX Software. Nutzt die vorgegebene Datei. Ihr könnt die Funktion mit der Maus bewegen und stauchen beziehungsweise strecken.
Hilfsmittel Schieberegler: Geogebra 5. Nutzt die bereitgestellte Geogebra-Datei. Ihr könnt die Funktion mit Hilfe der Schieberegler verändern.

Figure B.7 German Original: Minimally Guided Approach Including Scaffolding for all Four Groups

B.2.3   German Tasks Intermediate Approach

Ihr habt die Funktion $f(x) = x^2$ und ihren Graphen kennen gelernt. Zur Erinnerung seht ihr hier nochmal den Graphen.

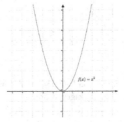

Man kann den Term durch das Hinzufügen von Faktoren oder Summanden verändern. Zum Beispiel gehört zum Term $f(x) = 4 \cdot (x - 2)^2 + 3$ der folgende Graph.

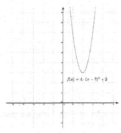

Aufgabe 1:
Überlegt euch in Partnerarbeit, was sich am Graphen der Funktion $f(x) = a \cdot (x - b)^2 + c$ verändert, wenn ihr a, b und c verändert. Vielleicht hilft es euch, wenn ihr zunächst nur a verändert, dann b und dann c. Versucht die Funktionen zu zeichnen.

Betrachtet verschiedene Beispiele und die zugehörigen Graphen.

Aufgabe 2: Versucht zu begründen, warum a, b oder c die jeweilige Veränderung am Graphen bewirkt.
Tipp: Betrachtet die Wertetabelle der Funktion und was eine Veränderung von a, b oder c bewirkt.

Aufgabe 3:
Erstellt zu zweit eine Art Spickzettel auf einem DIN A4 Blatt, auf dem ihr mit Begründung beschreibt, wie sich der Graph verändert. Der Spickzettel sollte so gestaltet werden, dass auch jemand zum Beispiel aus eurer Parallelklasse ihn versteht.

Figure B.8 German Original: *Without Visualization* Group in the Intermediate Approach

Ihr habt die Funktion $f(x) = x^2$ und ihren Graphen kennen gelernt. Zur Erinnerung seht ihr hier nochmal den Graphen.

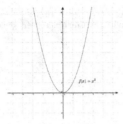

Man kann den Term durch das Hinzufügen von Faktoren oder Summanden verändern. Zum Beispiel gehört zum Term $f(x) = 4 \cdot (x-2)^2 + 3$ der folgende Graph.

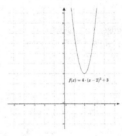

Aufgabe 1:
Überlegt euch in Partnerarbeit, was sich am Graphen der Funktion $f(x) = a \cdot (x-b)^2 + c$ verändert, wenn ihr a, b und c verändert. Vielleicht hilft es euch, wenn ihr zunächst nur a verändert, dann b und dann c. Versucht die Funktionen zu zeichnen.

Betrachtet verschiedene Beispiele und zeichnet ihre Graphen mit dem Rechner.

Aufgabe 2: Versucht zu begründen, warum a, b oder c die jeweiligen Veränderung am Graphen bewirkt.
Tipp: Betrachtet die Wertetabelle der Funktion und was eine Veränderung von a, b oder c bewirkt.
Die Wertetabelle der gezeichneten Funktionen könnt ihr euch die Werkzeugpalette Dokumente bei Punkt 7: Tabelle anzeigen lassen.

Aufgabe 3:
Erstellt zu zweit eine Art Spickzettel auf einem DIN A4 Blatt, auf dem ihr mit Begründung beschreibt, wie sich der Graph verändert. Der Spickzettel sollte so gestaltet werden, dass auch jemand zum Beispiel aus eurer Parallelklasse ihn versteht.

Figure B.9 German Original: *Function Plotter* Group in the Intermediate Approach

Ihr habt die Funktion $f(x) = x^2$ und ihren Graphen kennen gelernt. Zur Erinnerung seht ihr hier nochmal den Graphen.

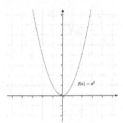

Man kann den Term durch das Hinzufügen von Faktoren oder Summanden verändern. Zum Beispiel gehört zum Term $f(x) = 4 \cdot (x - 2)^2 + 3$ der folgende Graph.

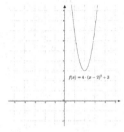

Aufgabe 1:
Überlegt euch in Partnerarbeit, was sich am Graphen der Funktion $f(x) = a \cdot (x - b)^2 + c$ verändert, wenn ihr a, b und c verändert. Vielleicht hilft es euch, wenn ihr zunächst nur a verändert, dann b und dann c. Versucht die Funktionen zu zeichnen.

Wenn ihr mit dem Cursor auf den Graphen geht und ihn „anpackt", könnt ihr den Graphen bewegen und strecken oder stauchen. Betrachtet auf diese Weise verschiedene Beispiele und die zugehörigen Graphen.

Aufgabe 2: Versucht zu begründen, warum a, b oder c die jeweiligen Veränderung am Graphen bewirkt.
Tipp: Betrachtet die Wertetabelle der Funktion und was eine Veränderung von a, b oder c bewirkt.

Aufgabe 3:
Erstellt zu zweit eine Art Spickzettel auf einem DIN A4 Blatt, auf dem ihr mit Begründung beschreibt, wie sich der Graph verändert. Der Spickzettel sollte so gestaltet werden, dass auch jemand zum Beispiel aus eurer Parallelklasse ihn versteht.

Figure B.10 German Original: *Drag Mode* Group in the Intermediate Approach

Ihr habt die Funktion $f(x) = x^2$ und ihren Graphen kennen gelernt. Zur Erinnerung seht ihr hier nochmal den Graphen.

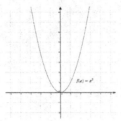

Man kann den Term durch das Hinzufügen von Faktoren oder Summanden verändern. Zum Beispiel gehört zum Term $f(x) = 4 \cdot (x - 2)^2 + 3$ der folgende Graph.

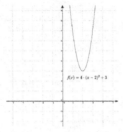

Aufgabe 1:
Überlegt euch in Partnerarbeit, was sich am Graphen der Funktion $f(x) = a \cdot (x - b)^2 + c$ verändert, wenn ihr a, b und c verändert. Vielleicht hilft es euch, wenn ihr zunächst nur a verändert, dann b und dann c. Versucht die Funktionen zu zeichnen.

Betrachtet verschiedene Beispiele und die zugehörigen Graphen mit dem Rechner in dem ihr die Funktion mit Hilfe der Schieberegler verändert.

Aufgabe 2: Versucht zu begründen, warum a, b oder c die jeweiligen Veränderung am Graphen bewirkt.
Tipp: Betrachtet die Wertetabelle der Funktion und was eine Veränderung von a, b oder c bewirkt.

Aufgabe 3:
Erstellt zu zweit eine Art Spickzettel auf einem DIN A4 Blatt, auf dem ihr mit Begründung beschreibt, wie sich der Graph verändert. Der Spickzettel sollte so gestaltet werden, dass auch jemand zum Beispiel aus eurer Parallelklasse ihn versteht.

Figure B.11 German Original: *Sliders* Group in the Intermediate Approach

Präsenzaufgabe 1 (Expertengruppen): Lösen Sie die untenstehende Aufgabe für die Klasse 9 in Partnerarbeit. Sie werden dazu in vier Gruppen aufgeteilt und lösen diese Aufgabe auf verschiedene Weise. Anschließend wollen wir mit Ihnen in einem Gruppenpuzzle über die verschiedenen Lösewege diskutieren. Überlegen Sie sich bereits in der Partnerarbeit, was Sie gut oder schlecht finden und wie man die Aufgabe verbessern könnte.

Figure B.12 German Original: Experts Group in the Testing of the Intermediate Approach

B.2.4   German Task Final Guided Discovery Approach

Dieses Arbeitsblatt soll euch dabei helfen, neue Erkenntnisse von quadratischen Funktionen eigenständig zu gewinnen. Diese sollt ihr am Ende in einem Merkblatt zusammenfassen.

1. Arbeitsauftrag: BEISPIEL BESCHREIBEN

Ihr habt die Funktion $f(x) = x^2$ und ihren Graphen kennen gelernt. Durch Verändern von Faktoren oder Summanden im Term verändert sich der Graph. Zum Beispiel $f(x) = 4 \cdot (x-2)^2 + 3$.

Beschreibt, wie sich der rechte Graph gegenüber dem linken Graph verändert hat.

2. Arbeitsauftrag: GRAPHEN UNTERSUCHEN

Untersucht, wie sich der Graph der Funktion $f(x) = a \cdot (x-b)^2 + c$ verändert, wenn ihr einen anderen Faktor (a) und andere Summanden (b und c) nutzt. Folgende Schritte können euch helfen:

1. Betrachtet zunächst c.
   Setzt in die Funktion $f(x) = x^2 + c$ verschiedene Werte für c ein. Zum Beispiel $-5, -2, -1, 0, 1, 2, 5$. Beobachtet, was passiert.

2. Betrachtet dann a.
   Setzt in die Funktion $f(x) = a \cdot x^2$ verschiedene Werte für a ein. Zum Beispiel $-5, -1, -\frac{1}{2}, 0, \frac{1}{2}, 1, 5$. Beobachtet, was passiert.

3. Betrachtet als letztes b.
   Setzt in die Funktion $f(x) = (x-b)^2$ verschiedene Werte für b ein. Zum Beispiel $-5, -2, -1, 0, 1, 2, 5$. Beobachtet, was passiert.

Nutzt dafür die vorgegebenen Medien. Bei Schwierigkeiten bei der Bedienung schaut ihr zunächst auf dem Merkblatt nach.

Figure B.13 German Original: Part One and Two of the Guided Discovery Approach

3. Arbeitsauftrag: BEGRÜNDUNGEN FINDEN

Im Folgenden sollt ihr nun versuchen, eure Ergebnisse aus Arbeitsauftrag 2 zu begründen. Folgende Schritte können euch helfen:

1. Begründet zunächst den Einfluss von c.

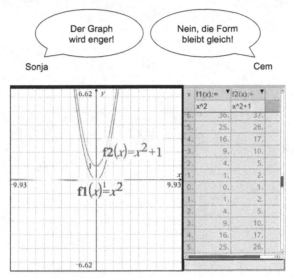

Diskutiert für die Begründung wer von beiden Recht hat und warum?

2. Begründet dann den Einfluss von a.
   Betrachtet dafür die Beispiele aus dem 2. Arbeitsauftrag und vergleicht jeweils mit der Normalparabel. Gib genau an, wie und warum sich der Graph verändert.

3. Begründet als letztes den Einfluss von b.
   Zeichnet dafür in ein Koordinatensystem die Graphen der Funktionen $f1(x) = x^2$ und $f2(x) = (x-1)^2$ und vollziehe die folgenden Aussagen nach.

Überlegt, wie ihr es noch begründen könnt.

4. Arbeitsauftrag: MERKBLATT ERSTELLEN

Erstellt nun auf einem DIN A4 Blatt euer Merkblatt mit allen Erkenntnissen, Ergebnissen und Begründungen.

Figure B.14 German Original: Part Three and Four of the Guided Discovery Approach

Full Coding Manual for Summary Sheet Analysis

In this chapter a translation of the coding manual for the summary sheet analysis can be found

© The Editor(s) (if applicable) and The Author(s), under exclusive license to Springer   241
Fachmedien Wiesbaden GmbH, part of Springer Nature 2021
L. Göbel, *Technology-Assisted Guided Discovery to Support Learning*,
Essener Beiträge zur Mathematikdidaktik,
https://doi.org/10.1007/978-3-658-32637-1

Content		
IGAS	Statements of the students mostly	1 viable
		2 non-viable
IUEA	Not needed statements/drawings (hearts, cats)	0 Not written
		1 Yes
INPA	Students mention parameter a	0 Not written
		1 Yes
INPB	Students mention parameter b	0 Not written
		1 Yes
INPC	Students mention parameter c	0 Not written
		1 Yes

Influence of a			
IAEB	Students recognize the influence of a	Vertical shrink	0 Not written
			1 viable
			2 non-viable
IAEE		Vertical stretch	0 Not written
			1 viable
			2 non-viable
IAEO		Reflection	0 Not written
			1 viable
			2 non-viable
IAEN		Special case a=0	0 Not written
			1 viable
			2 non-viable
IABB	Students explain the influence of a	Vertical shrink	0 Not written
			1 viable
			2 non-viable
IABE		Vertical stretch	0 Not written
			1 viable
			2 non-viable
IABO		Reflection	0 Not written
			1 viable
			2 non-viable
IABN		Special case a=0	0 Not written
			1 viable
			2 non-viable

Influence of b			
IBER	Students recognize influence of b	right	0 Not written
			1 viable
			2 non-viable
IBEL		left	0 Not written
			1 viable
			2 non-viable
IBEX		horizontal	0 Not written
			1 viable
			2 non-viable
IBBR	Students explain influence of b	right	0 Not written
			1 viable
			2 non-viable
IBBL		left	0 Not written
			1 viable
			2 non-viable
IBBX		horizontal	0 Not written
			1 viable
			2 non-viable

Influence of c			
ICEO	Students recognize influence of c	upwards	0 Not written
			1 viable
			2 non-viable
ICEU		downwards	0 Not written
			1 viable
			2 non-viable
ICEY		vertical	0 Not written
			1 viable
			2 non-viable
ICBO	Students explain influence of x	upwards	0 Not written
			1 viable
			2 non-viable
ICBU		downwards	0 Not written
			1 viable
			2 non-viable
ICBY		vertical	0 Not written
			1 viable
			2 non-viable

Representations			
DMKB	Students use the appropriate mathematical conventions		0 Not written
			1 Yes
Tabular representations			
DTDV	Tables used		0 Not written
			1 Yes
DTVV	Tables used as	illustration	0 Not written
			1 Yes, viable
			2 Yes, non-viable
DTVE		explanation	0 Not written
			1 Yes, viable
			2 Yes, non-viable
DTVB		example	0 Not written
			1 Yes, viable
			2 Yes, non-viable
DTVS		other	0 Not written
			1 Yes, viable
			2 Yes, non-viable

Graphical representation			
DGDV	Graphical representation used		0 Not written
			1 Yes
DGVV	Graphical representation used	illustration	0 Not written
			1 Yes, viable
			2 Yes, non-viable
DGVE		explanation	0 Not written
			1 Yes, viable
			2 Yes, non-viable
DGVB		example	0 Not written
			1 Yes, viable
			2 Yes, non-viable
DGVS		other	0 Not written
			1 Yes, viable
			2 Yes, non-viable

Symbolic Representation			
DSDV	Symbolic representation used		0 Not written
			1 Yes
DSVV	Symbolic representation used as	illustration	0 Not written
			1 Yes, viable
			2 Yes, non-viable
DSVE		explanation	0 Not written
			1 Yes, viable
			2 Yes, non-viable
DSVB		example	0 Not written
			1 Yes, viable
			2 Yes, non-viable
DSVS		other	0 Not written
			1 Yes, viable
			2 Yes, non-viable
Use of different representations			
DAUV	Different representations are connected		0 Not written
			1 viable
			2 non-viable

Structure			
SMST	Summary sheet is structured		1 very structured
			2 rather structured
			3 rather unstructured
			4 very unstructured
SMAS	Students use the structure of the worksheet to structure their summary sheet		0 restructured
			1 answers to task
			9 non decidable
SMUE	The summary sheet contains a title		0 Not written
			1 Yes
			9 non decidable
SMFO	Students writing		0 non existent
			1 only bullet points
			2 mostly bullet points
			3 mostly full sentences
			4 only full sentences
SHFA	Students highlight important things (content and structure)	Colours	0 Not written
			1 viable
			2 non-viable
SHUN		Underline	0 Not written
			1 viable
			2 non-viable
SHSH		Other highlight	0 Not written
			1 viable
			2 non-viable
SMAW	Word count		1 0-50
			2 50-100
			3 100-150
			4 150-200
			5 more than 200

Language use			
FGSP	What technical terms were mentioned	Vertex point	0 Not written
			1 Yes, viable
			2 Yes, non-viable
FGPA		Parabola	0 Not written
			1 Yes, viable
			2 Yes, non-viable
FGYA		y-axis	0 Not written
			1 Yes, viable
			2 Yes, non-viable
FGXA		x-axis	0 Not written
			1 Yes, viable
			2 Yes, non-viable
FGPO		Positive	0 Not written
			1 Yes, viable
			2 Yes, non-viable
FGNE		Negative	0 Not written
			1 Yes, viable
			2 Yes, non-viable
FGGA		„Squashed"	0 Not written
			1 Yes, viable
			2 Yes, non-viable
FGGE		„stretched"	0 Not written
			1 Yes, viable
			2 Yes, non-viable
FGGR		Graph	0 Not written
			1 Yes, viable
			2 Yes, non-viable
FGQF		Quadratic function	0 Not written
			1 Yes, viable
			2 Yes, non-viable
FGST		Gradient	0 Not written
			1 Yes, viable
			2 Yes, non-viable
FGRA		Up, down, left, right	0 Not written
			1 Yes, viable
			2 Yes, non-viable
FGAX		x-intercept	0 Not written
			1 Yes, viable
			2 Yes, non-viable
FGAY		y-intercept	0 Not written
			1 Yes, viable
			2 Yes, non-viable

FGSD		Slope triangle	0 Not written
			1 Yes, viable
			2 Yes, non-viable
FGNP		Standard parabola	0 Not written
			1 Yes, viable
			2 Yes, non-viable
FGKS		Coordinate system	0 Not written
			1 Yes, viable
			2 Yes, non-viable
FGSU		Summands	0 Not written
			1 Yes, viable
			2 Yes, non-viable
FGFA		Factor	0 Not written
			1 Yes, viable
			2 Yes, non-viable
FGFU		Function	0 Not written
			1 Yes, viable
			2 Yes, non-viable
FGFO		Formula	0 Not written
			1 Yes, viable
			2 Yes, non-viable
FGKO		Coordinate	0 Not written
			1 Yes, viable
			2 Yes, non-viable
FGAD		Add	0 Not written
			1 Yes, viable
			2 Yes, non-viable
FGSB		Subtract	0 Not written
			1 Yes, viable
			2 Yes, non-viable
FGMU		Multiplicate	0 Not written
			1 Yes, viable
			2 Yes, non-viable
FGTE		Term	0 Not written
			1 Yes, viable
			2 Yes, non-viable
FGWT		Function tables	0 Not written
			1 Yes, viable
			2 Yes, non-viable
FGQU		Quadrant	0 Not written
			1 Yes, viable
			2 Yes, non-viable
FGSA		Symmetry axis	0 Not written
			1 Yes, viable
			2 Yes, non-viable

FGAS		Axis-symmetrical	0 Not written
			1 Yes, viable
			2 Yes, non-viable
FGSO		Other technical terms	0 Not written
			1 Yes, viable
			2 Yes, non-viable
FKGF	Same term used for the same concept	Parabola and quadratic Function	0 only parabola
			1 only quadratic Function
			2 both
FAFB	Number of technical terms used		0 none
			1: 1-5
			2: 6-10
			3: >10
FVAF	Use of everyday and professional language		0 only professional language
			1 mix of everyday and professional language, everyday language used for explanation/illustration
			2 Switch between everyday and professional language, because of lack of technical terms
			3 only everyday language use

Bilingual Examples of Summary Sheet Categories

D

D.1 Parameter a

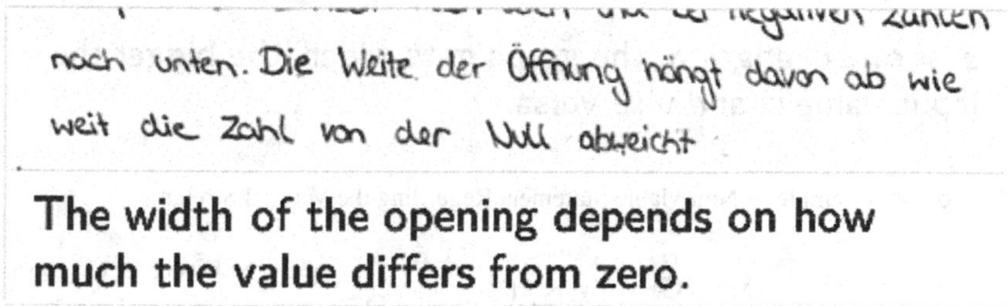

noch unten. Die Weite der Öffnung hängt davon ab wie weit die Zahl von der Null abweicht.

The width of the opening depends on how much the value differs from zero.

Figure D.1 Example of Viable Statement Regarding the Vertical Shrink

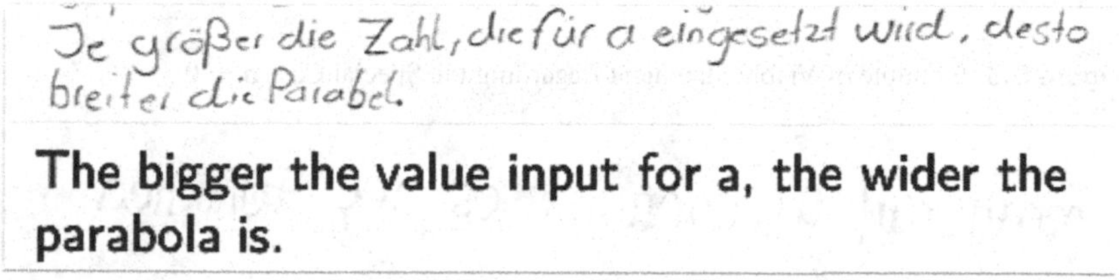

Je größer die Zahl, die für a eingesetzt wird, desto breiter die Parabel.

The bigger the value input for a, the wider the parabola is.

Figure D.2 Example of Non-viable Statement Regarding the Vertical Shrink

© The Editor(s) (if applicable) and The Author(s), under exclusive license to Springer Fachmedien Wiesbaden GmbH, part of Springer Nature 2021
L. Göbel, *Technology-Assisted Guided Discovery to Support Learning*,
Essener Beiträge zur Mathematikdidaktik,
https://doi.org/10.1007/978-3-658-32637-1

> Wenn man für a eine positive Zahl einsetzt, wird die Parabel von kleinen zu großen Zahlen immer dünner.

If one sets for a a positive value, the parabola gets narrower from small to big values.

Figure D.3 Example of Viable Statement Regarding the Vertical Stretch

> a. Wenn man a verändert wird der Graph breiter desto höher die eingesetzen Zahlen sind genauso andersrum

a. If one changes a, the graph gets wider, the bigger the input value is and vice versa.

Figure D.4 Example of Non-viable Statement Regarding the Vertical Stretch

> die Breite des Graphen. Wird für a → 0 eingesetzt Liegt der Graph auf oder parallel der x-Achse.

If for a -> 0 is input, the graph lies on or parallel to the x-axis.

Figure D.5 Example of Viable Statement Regarding the Special Case $a = 0$

> genau auf 0: dann ist es eine Funktion

exactly on 0: then it is a function

Figure D.6 Example of Non-viable Statement Regarding the Special Case $a = 0$

Wird a verändert, z.B. eine negative wird eingesetzt,
so öffnet sich die Parabel nach unten und der
Scheitelpunkt ist oben.

**If a is changed, e.g. a negative value is input,
the parabola opens downwards and the vertex
point is at the top.**

Figure D.7 Example of Viable Statement Regarding the Reflection of the Parabola

verändert man (a), so verändert sich die Parabel nur in der Höhe
nicht in der Breite.

**If one changes (a), the parabola only changes
in height not in width.**

Figure D.8 Example of Non-viable Statement Regarding the Reflection of the Parabola

D.2   Parameter b

b bestimmt die Position auf der x-Achse.
(bei Minus nach links, bei plus nach rechts)

**b determines the position on the x-axis.
(with minus to the left, with plus to the right).**

Figure D.9 Example of Viable Statement Regarding the Movement to the Left and Right

$\Rightarrow$ b bedeutet

Es verschiebt den Scheitelpunkt ($\rightarrow$ Parabel)
+b nach rechts und -b nach links.

$\Rightarrow$ b represents
It moves the vertex (-> Parabola)
+b to the right and −b to the left.

Figure D.10 Example of Non-viable Statement Regarding the Movement to the Left and Right as Well as the Horizontal Transformation

Wenn man die Formel $f(x)=(x-b)^2$ hat und b verändert, verändert sich die Position auf der x-Achse.

If one has the formula $f(x) = (x - b)^2$ and changes b, the position on the x-axis changes.

Figure D.11 Example of Viable Statement Regarding the Horizontal Transformation

- „b" bestimmt, in welchem Quadrant der Scheitelpunkt liegt

- „b" determines in which quadrant the vertex lies

Figure D.12 Example of Non-viable Statement Regarding the Horizontal Transformation

D.3   Parameter c

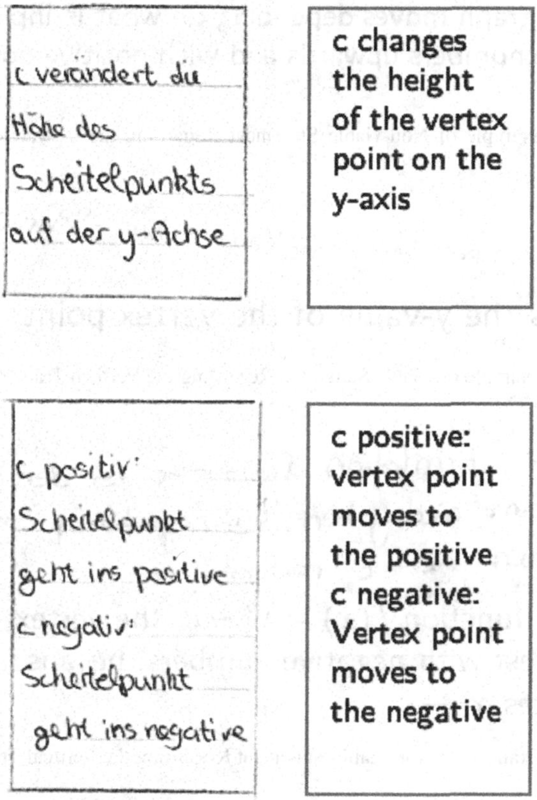

c verändert die Höhe des Scheitelpunkts auf der y-Achse	c changes the height of the vertex point on the y-axis
c positiv: Scheitelpunkt geht ins positive c negativ: Scheitelpunkt geht ins negative	c positive: vertex point moves to the positive c negative: Vertex point moves to the negative

Figure D.13 Example of Viable Statement Regarding the Upwards and Downwards Movement

1.) Der Graph verschiebt sich jenachdem was man eingibt bei negativen Zahlen nach oben und bei positiven nach unten.

1.) The graph moves depending on what is input, with negative numbers upwards and with positive downwards.

Figure D.14 Example of Non-viable Statement Regarding the Upwards and Downwards Movement

c ≠   c ist der y-Wert des Scheitelpunkts

c= c is the y-value of the vertex point

Figure D.15 Example of Viable Statement Regarding the Vertical Transformation

Bei der Funktion $f(x) = x^2 - c$ ist der Scheitel-punkt am niedrigsten, bei negativen Zahlen weil man $x^2 - c$ rechnet.

For the function $f(x) = x^2 - c$, the vertex point is the lowest with negative numbers, because one calculates $x^2 - c$.

Figure D.16 Example of Non-viable Statement Regarding the Vertical Transformation

D.4   Language Use

b verändert die Höhe des Scheitelpunkts auf der x-Achse.	b changes the height of the vertex point on the x-axis.
c verändert die Höhe des Scheitelpunkts auf der y-Achse	c changes the height of the vertex point on the y-axis

Figure D.17 Example of Viable Use of *x-axis* (Left) and *y-axis* (Right)

Figure D.18 Example of Viable Use of *Slope*

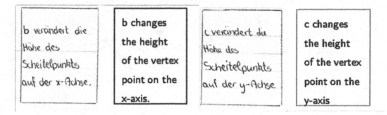

3.2 α beinflusst die Steigerung

3.2 a influences the slope

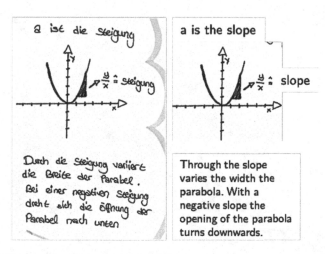

a ist die Steigung

$\frac{y}{x} \triangleq$ Steigung

Durch die Steigung variiert die Breite der Parabel. Bei einer negativen Steigung dreht sich die Öffnung der Parabel nach unten

a is the slope

$\frac{y}{x} \triangleq$ slope

Through the slope varies the width the parabola. With a negative slope the opening of the parabola turns downwards.

Figure D.19 Example of Non-viable Use of *Slope*

Comparison Summary Sheets Between Dynamic and Static Groups

Table E.1 Results Parameter *a* Split Between Static and Dynamic Groups

Category			Static Groups (N = 138)	Dynamic Groups (N = 215)	Total (N = 353)
Vertical Shrink	$p < 0.001$	Not written	85 (61.6%)	79 (36.7%)	164 (46.5%)
		Viable	40 (29.0%)	109 (50.7%)	149 (42.2%)
		Non-Viable	13 (9.4%)	27 (12.6%)	40 (11.3%)
Vertical Stretch	$p < 0.001$	Not written	53 (38.4%)	43 (20.0%)	96 (27.2%)
		Viable	75 (54.3%)	137 (63.7%)	212 (60.1%)
		Non-Viable	10 (7.2%)	35 (16.3%)	45 (12.7%)
Reflection	$p < 0.001$	Not written	82 (59.4%)	73 (34.0%)	155 (43.9%)
		Viable	54 (39.1%)	126 (58.6%)	180 (51.0%)
		Non-Viable	2 (1.4%)	16 (7.4%)	18 (5.1%)
Case a = 0	$p < 0.001$	Not written	138 (100%)	178 (82.8%)	316 (89.5%)
		Viable	0 (0%)	37 (17.2%)	37 (10.5%)
		Non-Viable	0 (0%)	0 (0%)	0 (0%)

For all three parameters in all categories, Chi-Square tests show statistically significant differences between the static, so *without visualization* and *function plotter* groups and the dynamic, so *drag mode* and *sliders* group.

L. Göbel, *Technology-Assisted Guided Discovery to Support Learning*, Essener Beiträge zur Mathematikdidaktik, https://doi.org/10.1007/978-3-658-32637-1

Table E.2 Results Parameter b Split Between Static and Dynamic Groups

Category			Static Groups (N = 138)	Dynamic Groups (N = 215)	Total (N = 353)
Right	$p = 0.001$	Not written	100 (72.5%)	114 (53.0%)	214 (60.6%)
		Viable	23 (16.7%)	73 (34.0%)	96 (27.2%)
		Non-Viable	15 (10.9%)	28 (13.0%)	43 (12.2%)
Left	$p < 0.001$	Not written	104 (75.4%)	117 (54.4%)	221 (62.6%)
		Viable	19 (13.8%)	70 (32.6%)	89 (25.2%)
		Non-Viable	15 (10.9%)	28 (13.0%)	43 (12.2%)
Horizontal	$p < 0.001$	Not written	34 (24.6%)	15 (7.0%)	49 (13.9%)
		Viable	79 (57.2%)	162 (75.3%)	241 (68.3%)
		Non-Viable	25 (18.1%)	38 (17.7%)	63 (17.8%)

Table E.3 Results Parameter c Split Between Static and Dynamic Groups

Category			Static Groups (N = 138)	Dynamic Groups (N = 215)	Total (N = 353)
Upwards	$p = 0.003$	Not written	97 (70.3%)	112 (52.1%)	209 (59.2%)
		Viable	37 (26.8%)	94 (43.7%)	131 (37.1%)
		Non-Viable	4 (2.9%)	9 (4.2%)	13 (3.7%)
Downwards	$p = 0.002$	Not written	103 (74.6%)	123 (57.2%)	226 (64.0%)
		Viable	33 (23.9%)	81 (37.7%)	114 (32.3%)
		Non-Viable	2 (1.4%)	11 (5.1%)	13 (3.7%)
Vertical	$p < 0.001$	Not written	20 (14.5%)	7 (3.3%)	27 (7.6%)
		Viable	102 (73.9%)	185 (86.0%)	287 (81.3%)
		Non-Viable	16 (11.6%)	23 (10.7%)	39 (11.0%)

Description of the Two Examples of Guided Discovery

F

In this chapter the two processes described in chapter 12 and especially through Figures 12.12 and 12.5 will be described in more detail. The number corresponds to the number in the phase rectangular in the graphic timelines.

F.1 The Process of Charlotte and Noah Using Function Plotters

1. Charlotte and Noah begin by organizing their work and put their student codes on their papers (Minutes 0–1:51).
2. They then start reading the first task on the worksheet (1:51–2:23).
3. Following the task they describe the differences between the two graphs $f(x) = x^2$ and $f(x) = 4 \cdot (x - 2)^2 + 3$ in a graphical manner and conclude that the transformed parabola is smaller and moved to the right (2:23–3:03).
4. They refer the differences to the equation and discuss, what the y-intercept is. During this discussion, Charlotte shows parameter stereotyping, when identifying b as the y-intercept (3:03–3:52).
5. They then recap their results in order to write them down (3:52–4:53). *Insights: The graph is moved and smaller than the standard parabola.*
6. While working on task 2.1 (For c in $f(x) = x^2 + c$ use different values, for example $-5, -2, -1, 0, 1, 2, 5$. What changes then?) the students show missing technological knowledge as they do not know how to input the x^2. They resolve this by asking the teacher how to input it. The technological obstacle hinders an insight (4:53–6:35).
7. After Charlotte and Noah overcome the first technological obstacle, they encounter another one as they do not know, how to graph the function they input. They press the necessary input button by chance and succeed in graph-

259

L. Göbel, *Technology-Assisted Guided Discovery to Support Learning*, Essener Beiträge zur Mathematikdidaktik, https://doi.org/10.1007/978-3-658-32637-1

ing the first function $f(x) = x^2 + (-5)$. Noah zooms in and out and consults the task again (6:35–7:59).

8. They input a further equation ($f(x) = x^2 + (-2)$) and describe the differences. They choose to graph another equation, but before doing this, they state the hypothesis that it would move upwards by one. Then, they proceed to plot another function, however hey forget the plus sign between x^2 and -1 so a reflected parabola is shown. Noah realised their mistake and they correctly input $x^2 + (-1)$. Noah describes the influence of c, but conjectures that the graph gets narrower. He then inputs $f(x) = x^2 + 5$ and Charlotte presumes that c is the y-intercept. *Insights: After use of technology: Parabola moves upwards for bigger values, c is the y-intercept* (7:59–9:36)

9. Noah and Charlotte try to reset the app to the starting position again. As they do not succeed with deleting the graphs, they press the undo button until they reach the starting point again (9:36–10:25).

10. After the starting point is re-established, the students go to task 2.2 (Now look at a. Use different values of a in $f(x) = a \cdot x^2$ for example $-5, -1, -\frac{1}{2}, 0, \frac{1}{2}, 1, 5$. Watch what happens). They input $f(x) = -5 \cdot x^2$ and conclude that the graph is reflected and hypothesise that it is narrow and would be wider, when another example is graphed. They then test this by graphing $f(x) = -2 \cdot x^2$ (10:25–10:59).

11. Planning to graph $f(x) = \frac{1}{2} \cdot x^2$, they face another technical obstacle, as they do not know how to input fractions. This obstacle hinders an insight, as they do not overcome the difficulty, but rather choose a different value for a. (10:59–11:35)

12. Charlotte refers the change in the graphs to the formula and identifies a as the slope. She argues correctly that if a is negative the parabola is opened downwards, all the while showing parameter stereotypes referring to a as m (which is often used as the parameter for slope in linear functions). *Insight:* **Non-viable** *insight that a is the exact slope.* (11:35–12:27)

13. Charlotte and Noah proceed to task 2.3 (Finally look at b. Use different values of b in $f(x) = (x - b)^2$ for example $-5, -2, -1, 0, 1, 2, 5$. Watch what happens.) and first restore the blank page in the TI-Nspire CX CAS app. They input $f(x) = (x - 5)^2$ and conjecture that b is the position on the x-axis. Noah proposes to test this with another example and they input $f(x) = (x - 2)^2$ as confirmation and as an example for negative values of b they input $f(x) = (x - (-2))^2$ as another confirmation of their hypothesis. *Insight: b is the position of the parabola on the x-axis.* (12:27–14:22).

14. Wishing to write down the key points for each parameter, Charlotte recapitulates the insights for c again trying to match the parameters of quadratic functions to the ones learnt with linear functions. She identifies c with the b

of linear functions of the kind $y = m \cdot x + b$ and therefore states that c is the y-intercept, which is correct for all examples they graphed. Both Noah and Charlotte state the misconception that the graph is narrower if c is changed. Charlotte recognises a cognitive conflict as she correctly assumes that the slope does not change. (14:22–17:23)

15. They move to the key points for parameter a and describe the influence of a correctly, Charlotte then poses the question what happens if $a = 0$. (17:23–18:32)

16. Before testing what happens if $a = 0$, they proceed to recapitulate the insights for parameter b. (18:32–19:03)

17. On Charlotte's request they input then $f(x) = 0 \cdot x^2$ and Charlotte reacts to the graph on the x-axis with no surprise. Noah inputs one of the equations graphed earlier again ($f(x) = -5 \cdot x^2$) as confirmation of their insights. Charlotte describes the changes in the graph dynamically ("gets wider, with zero its on the x-axis and that it goes upwards") and Noah tests this with an example not given in the task and plots $f(x) = 6 \cdot x^2$. Charlotte tries describing the influence, but does not know the correct technical term for parameters and uses variable instead. *Insight: Identification of the special case $a = 0$ and description of influence through a.* (19:03–22:40)

18. Noah reads out the task 3 (Now try to find an explanation for your findings in exercise 2. The following steps may be helpful again: Try to explain the influence of c) and thinks that they have already done what is asked of them. Charlotte suggests skipping task three and doing task four first. (22:40–23:44)

19. Noah then reads the two statements of task 3.1 (The graph gets narrower. No, it stays the same! Discuss with your neighbour if Sonja or Cem is right.) and they discuss the statements. They do not see the conflict between the two statements and Charlotte agrees with Sonja that the graph is narrower. (23:44–24:32)

20. While working on task 3.2 (Now try to explain the influence of a. Look at your findings in exercise 2 again and compare to the graph of $f(x) = x^2$. Specify exactly how and why the graph changes.), both Charlotte and Noah only restate their insight and do not find explanations for the change. (24:32–25:41)

21. For the last part of task 3 (Finally try to explain the influence of b. Sketch the graphs of $f1(x) = x^2$ and $f2(x) = (x - 1)^2$ in one coordinate system.) they graph the two functions and try to understand the two explanations given. During this discussion they restate that changing b moves the graph on the x-axis and due to her parameter stereotypes Charlotte is confused by b as it is always the y-intercept for her. For the second given explanation they try to understand it by graphing $(3 - 1)^2$ but do not succeed in understanding the statements. (25:41–31:46)

22. Noah and Charlotte start planning their summary sheet. (31:46–32:34)
23. While working on task two, Noah wrote their findings down but numbered the tasks with a, b, c instead of 2.1, 2.2, 2.3. Due to this, when starting to write their summary sheet they are confused with the influences but overcome the confusion and write a statement regarding the influence of a. (32:34–34:31)
24. Following this, they write a statement about the influence of b. (34:31–35:13)
25. While discussing what to write for parameter c, Noah identifies the conflict in their earlier statements that the graph gets narrower. In order to check this he plots the graphs $f(x) = x^2 - 5$ and $f(x) = x^2 + 5$ and correctly explains, why the graph is not narrower, but only moved upwards using the technology. *Insight: Overcoming of the misconception that c changes shape of the graph.* (35:13–37:32)
26. They then assume they are finished with the task and tell the researcher this. (37:32–38:24)
27. The researcher asks them to explain their results and while they are doing this, the researcher tries to evoke explanations through posing questions. Even though Noah and Charlotte can argue the influence rather good, the explanations elude them. As they have not graphed an example where $|a| < 1$, the researcher asks them to do this. (38:24–42:24)
28. They proceed to graph $f(x) = 0.5 \cdot x^2$ and $f(x) = 5 \cdot x^2$ and show the researcher this. (42:24–42:48)
29. The researcher inputs $f(x) = x^2$ and encourages the students to write their insights through the comparison of the three graphs onto the summary sheet. (42:48–44:17)
30. Noah deletes the three graphs and inputs $f(x) = 5 \cdot x^2$, $f(x) = 0.5 \cdot x^2$ and $f(x) = x^2$. While trying to input $f(x) = x^2 \cdot -5$, Noah forgets the multiplication sign and the graph $f(x) = x^2 - 5$ is shown. This confounds them both, but they quickly spot their mistake and correct it to $f(x) = x^2 \cdot (-5)$. They then describe the reflection through a and identify that the parabolas for $f(x) = 5 \cdot x^2$ and $f(x) = -5 \cdot x^2$ are identical apart from the reflection. Noah chooses to input his own example with $f(x) = -10 \cdot x^2$ and describes it as narrower. *Insight: Negative values for a result in a reflection at the x-axis.* (44:17–47:06)
31. The researcher adds the function for $f(x) = -0.5 \cdot x^2$ and $f(x) = -x^2$ and tries to foster explanations again. The students do not succeed in this. (47:06–48:48)
32. In the last ten minutes of the intervention, the students try to formulate their insights for parameter a for the summary sheet. (48:48–58:44)

F.2   The Process of Tom and Iwan Using Sliders

1. The intervention starts with the researcher explaining the task for Tom and Iwan. (0:00–1:02)
2. They start with task one (Describe, how the right graph changed compared to the left graph.) and Tom decides to input the values of the changed graph $f(x) = 4 \cdot (x-2)^2 + 3$ with the sliders. As the sliders are hard to input the exact values, Tom uses the settings of the sliders to input the start value. (1:02–2:50)
3. They proceed to describe the changes between the two graphs, during this Iwan states that the transformed parabola does not have a y-intercept. Tom explains that he is wrong by using the technology to show this. (2:50–3:46)
4. Tom then starts to write notes on the influence of c. Tom is unsure if c is zero or one, if the parabola lies on the x-axis, so he checks this using the technology moving the slider for c to zero and then uses the settings to change it back to three. *Insight: Only if $c = 0$ the parabola lies on the x-axis.* (3:46–4:52)
5. They then discuss when a graph has its vertex point at the origin and if both parameter b and c have to be unequal to zero or if only one of it is enough. (4:52–5:45)
6. In order to test what value the parameter have to be they change the sliders for b and c to 0.1 and 0 respectively and then move the slider for b to four and back and identify the movement of the graph and then repeat this for slider c. *Insight: As soon as one of the parameters b or c is unequal to zero, the graph does not have its vertex point in the origin. Changing b moves the graph along the x-axis, while c influences the position on the y-axis.* (5:45–6:25)
7. Tom and Iwan then start formulating their insights for their notes and identify b as the x-intercept and c as the y-intercept and describe the movement of the parabola. (6:25–7:15)
8. They go back to task one and identify the change in the width due to the higher value of a. Tom changes the slider for a and describes the changes in the graph through exploration. While writing down their results Tom hypothesizes that the graph will be reflected if a is negative and then tests this with moving the slider for a to a negative value for the first time. *Insight: Influence of a on width of the parabola including special case $a = 0$ and reflection through negative values.* (7:15–8:42)
9. Elaborating on the insights regarding the influence of a, Tom describes what happens while he moves the slider for a. They try explaining their results with respect to the coordinates. (8:42–9:54)
10. They then decide to proceed to the second task of the intervention and realise that they already have done this and recap their results and then restore the starting

position of the sliders. Tom confirms that the shown graph is the standard parabola going through the values of the function table. (9:54–11:27)

11. The researcher prompts them to explore some examples even though they think they have already done the task, Tom and Iwan adjust the slider for c from -5 to -2 to finally -1 and recap that c is the y-intercept. However, Tom realises that c only gives the y-coordinate of the vertex point and not necessarily the y-intercept. *Insight: c is the y-coordinate of the vertex point.* (11:27–12:39)

12. They are briefly distracted with informal talk. (12:39–13:01)

13. The go back to the task and move the slider for a to -5 and Tom explains using the function table and mapping values, how one computes the y-value for a x-value. (13:01–14:37)

14. Iwan shows the misconception that x always has a value of one. He also tries to split the formula completely into x- and y-intercept. Tom explains that this is not accurate and how the y-value is computed. (14:37–16:13)

15. Iwan states his confusion and states that a changes both the x- and $y-$ intercept. (16:13–16:37)

16. To overcome the confusion, Tom moves the slider for b to -5 and explains that b is the x-coordinate of the vertex point. However, Iwan still states that x is always 1. Tom resolves this, by explaining what values x takes. (16:37–17:50)

17. They formulate their results for parameter b. *Insight: b is the x-coordinate of the vertex point.* (17:50–18:29)

18. Afterwards, they are briefly distracted through the technology. (18:29–19:05)

19. While reading task 2.1 Iwan thinks $f(x) = x^2 + c$ is a different equation to $f(x) = a \cdot (x - b)^2 + c$. This leads to the discussion during which Tom explains how the two equations are connected. (19:05–20:53)

20. They then proceed to task three (Finding explanations). Whilst they are discussing, which of the two statements is correct, Tom states the misconception that the graph gets narrower, but they quickly correct it after Iwan recaps their insights. (20:53–21:55)

21. For task 3.2 (Explanation of influence of a) they recap their results and Tom explains the influence using the mapping Grundvorstellung. (21:55–23:04)

22. Then they start working on task 3.2 and input the second function $f(x) = (x - 1)^2$. (23:04–25:07)

23. Iwan tries to argue using the function table, but both of them are confused with the position of the parabola. Tom changes the slider for b to -1 and wonders why the function they input has its vertex point at $x = 1$ and the one manipulated with sliders has its vertex point at $x = -1$. They input the second function again, but achieve the same results. Tom realises that if $b = -1$ the

function equation is $f(x) = (x - (-1))^2$ and checks this by altering the second function. (25:07–27:59)

24. Reconstructing the first statement of task 3.3, the students are still confused, as they are now comparing $f(x) = (x + 1)^2$ and $f(x) = x^2$, this leads to contradicting results in the function table. (27:59–30:26)

25. However, they try reasoning for the first statement with the values in the function table using the mapping Grundvorstellung. Iwan tries explaining the statement, while moving the second function with the drag mode parallel to the x-axis, he still has the misconception that $x = 1$. (30:26–33:05)

26. They choose to work with the second statement and try to reconstruct it with the function table. Iwan then realises that they mistakenly had the wrong value for b and they correct it. However, further explanations were not explored. (33:04–35:02)

27. Starting on their summary sheet they recap their results for the three parameter and write short statements for the influence. (35:02–37:21)

28. Elaborating on the influence of a, they describe the exact influences and include the special case $a = 0$, where the graph is a straight line on the x-axis and the reflection, if a is negative. (37:21–40:03)

29. Describing the influence of b, they identify that the parabola moves horizontally. (40:03–40:56)

30. Iwan mistakes c for the y-intercept, this leads to a discussion about the concept of the y-intercept. Tom explains the concept using the technology. (40:56–42:01)

31. After this discussion, they go back to formulating their summary sheet. (42:01–43:04)

32. Tom finalises the summary sheet with the recap of their insight regarding c and checks their insight using the technology. They finish work on their summary sheet with Tom reading it out and adding the function equations. (43:04–46:19)

33. The last four minutes of the intervention, they write their student codes down and then start informal talking. (46:19–50:06)

References

AINSWORTH, S. (1999). The functions of multiple representations. *Computers & Education*, *33*(2-3), 131–152.

AINSWORTH, S. (2006). DeFT: A conceptual framework for considering learning with multiple representations. *Learning and Instruction*, *16*(3), 183–198

AINSWORTH, S. (2014). The Multiple Representation Principle in Multimedia Learning. In R. MAYER (Ed.), *The Cambridge Handbook of Multimedia Learning* (pp. 464–486). Cambridge University Press.

AKINWUNMI, K. (2012). *Zur Entwicklung von Variablenkonzepten beim Verallgemeinern mathematischer Muster.*

ALFIERI, L., BROOKS, P. J., ALDRICH, N. J., & TENENBAUM, H. R. (2011). Does Discovery-Based Instruction Enhance Learning? *Journal of Educational Psychology*, *103*(1), 1–18.

ARCAVI, A., DRIJVERS, P., & STACEY, K. (2017). *The Learning and Teaching of Algebra – Ideas, Insights and Activities.* Abingdon, Oxon; New York, NY: Routledge.

ARNON, I., COTTRILL, J., DUBINSKY, E., OKTAÇ, A., ROA FUENTES, S., TRIGUEROS, M., & WELLER, K. (2014). *APOS Theory – A Framework for Research and Curriculum Development in Mathematics Education.* New York: Springer Science & Business Media LLC.

ARTIGUE, M. (2002). Learning Mathematics in a CAS Environment: The Genesis of a Reflection About Instrumentation and the Dialectics Between Technical and Conceptual Work. *International Journal of Computers for Mathematical Learning*, *7*(3), 245–274.

ASIALA, M., BROWN, A., DEVRIES, D. J., DUBINSKY, E., MATHEWS, D., & THOMAS, K. (1996). A framework for research and curriculum development in undergraduate mathematics education. In J. KAPUT, A. H. SCHOENFELD, & E. DUBINSKY (Eds.), *Research in collegiate mathematics education* (Vol. 2, pp. 1–32). Washington: AMS.

AUSUBEL, D. P. (1963). *The Psychology of Meaningful Verbal Learning.* New York: Grune & Stratton.

AUSUBEL, D. P. (1964). Some psychological and educational limitations of learning by discovery. *The Arithmetic Teacher*, *11*(5), 290–302.

AUSUBEL, D. P., NOVAK, J. D., & HANESIAN, H. (1978). *Educational Psychology: A Cognitive View* (2nd ed.). New York: Holt, Rinehart & Winston.

BAKER, B., HEMENWAY, C., & TRIGUEROS, M. (2000). On Transformations of Basic Functions. In H. L. CHICK, K. STACEY, & J. L. VINCENT (Eds.), *Proceedings of the 12th ICMI study conference on the future of the teaching and learning of algebra* (Vol. 1, pp. 41–47). University of Melbourne.

BAKER, B., HEMENWAY, C., & TRIGUEROS, M. (2001). On Transformations of Functions. In R. SPEISER, C. A. MAHER, & C. N. WALTER (Eds.), *Proceedings of the Annual Meeting of the North American Chapter of the International Group for the Psychology of Mathematics Education* (Vol. 2, pp. 91–98). Snowbird, Utah: International Group for the Psychology of Mathematics Education. North American Chapter.

BALL, L., & BARZEL, B. (2018). Communication When Learning and Teaching Mathematics with Technology. In L. BALL, P. DRIJVERS, S. LADEL, H.-S. SILLER, M. TABACH, & C. VALE (Eds.), *Uses of Technology in Primary and Secondary Mathematics Education* (Chap. 12, pp. 227–243). Cham: Springer International Publishing.

BARDINI, C., RADFORD, L., & SABENA, C. (2005). Struggling with variables, parameters, and indeterminate objects or how to go insane in mathematics. In H. L. CHICK & J. L. VINCENT (Eds.), *Proceedings of the 29th Conference of the International Group for the Psychology of Mathematics Education* (Vol. 2, pp. 129–136). Melbourne.

BAROODY, A. J., PURPURA, D. J., EILAND, M. D., & REID, E. E. (2015). The impact of highly and minimally guided discovery instruction on promoting the learning of reasoning strategies for basic add-1 and doubles combinations. *Early Childhood Research Quarterly, 30*, 93–105.

BARZEL, B. (2006). *Mathematikunterricht zwischen Konstruktion und Instruktion: Evaluation einer Lernwerkstatt im 11. Jahrgang mit integriertem Einsatz von Computeralgebra* (Doctoral dissertation, Fakultät für Mathematik, Universität Duisburg-Essen). Retrieved from http://duepublico.uni-duisburg-essen.de/servlets/DocumentServlet?id=13537

BARZEL, B. (2012). *Computeralgebra im Mathematikunterricht: Ein Mehrwert – aber wann?* Münster: Waxmann.

BARZEL, B., & GREEFRATH, G. (2015). Digitale Mathematikwerkzeuge sinnvoll integrieren. In W. BLUM, S. VOGEL, C. DRÜKE-NOE, & A. ROPPELT (Eds.), *Bildungsstandards aktuell: Mathematik in der Sekundarstufe II* (Chap. 2, pp. 145–157). Braunschweig: Diesterweg Schroedel Westermann.

BARZEL, B., & HOLZÄPFEL, L. (2017). Strukturen als Basis der Algebra. *Mathematik Lehren, 202*, 2–8.

BARZEL, B., HUSSMANN, S., & LEUDERS, T. (2005). *Computer, Internet & Co im Mathematikunterricht.* Berlin: Cornelsen Scriptor.

BILLS, L. (1997). Stereotypes of literal Symbol use in Senior School Algebra. In E. PEHKONEN (Ed.), *Proceedings of the Conference of the International Group for the Psychology of Mathematics Education* (Vol. 2, pp. 81–88).

BILLS, L. (2001). Shifts in the meanings of literal symbols. In M. VAN DEN HEUVEL-PANHUIZEN (Ed.), *Proceedings of the 25th Conference of the International Group for the Psychology of Mathematics Education* (Vol. 2, pp. 161–168). Freudenthal Institute. Utrecht, Netherlands.

BLOEDY-VINNER, H. (2001). Beyond unknowns and variables - parameters and dummy variables in high school algebra. In R. SUTHERLAND, T. ROJANO, A. BELL, & R. LINS (Eds.), *Perspectives on school algebra* (pp. 177–189). Dordrecht, Netherlands: Kluwer.

BLUM, W. (2004). On the role of "Grundvorstellungen" for reality-related proofs – Examples and reflections. In M. A. MARIOTTI (Ed.), *CERME-3 – Proceedings of the Third Conference of the European Society for Research in Mathematics Education*, Universitá di Pisa.

BOERS, M. A., & JONES, P. L. (1994). Students' use of graphics calculators under examination conditions. *International Journal of Mathematical Education in Science and Technology*, *25*(4), 491–516.

BORBA, M. C., & CONFREY, J. (1996). A Student's Construction of Transformations of Functions in a Multiple Representational Environment. *Educational Studies in Mathematics*, *31*, 319–337.

BORBA, M. d. C. (1993). *Students' understanding of transformations of functions using multirepresentational software* (Doctoral dissertation, Cornell University).

BORWEIN, J. M. (2005). The Experimental Mathematician: the Pleasure of Discovery and the Role of Proof. *International Journal of Computers for Mathematical Learning*, *10*(2), 75–108.

BORWEIN, J. M., & BAILEY, D. H. (2008). *Mathematics by Experiment: Plausible Reasoning in the 21st Century* (2nd ed.). CRC Press.

BREIDENBACH, D., DUBINSKY, E., HAWKS, J., & NICHOLS, D. (1992). Development of the process conception of function. *Educational Studies in Mathematics*, 23(3), 247–285.

BROWN, A. L., & CAMPIONE, J. C. (1994). Guided Discovery in a Community of Learners. In K. MCGILLY (Ed.), *Classroom Lessons: Integrating Cognitive Theory and Classroom Practice* (Chap. 9, pp. 229–270). Cambridge, Massachusetts, USA: The MIT Press.

BRUNER, J. S. (1961). The act of discovery. *Harvard Educational Review*, *31*(1), 21–32.

CAVANAGH, M., & MITCHELMORE, M. (2000a). Graphics calculators in mathematics learning: Studies of student and teacher understanding. In M. O. J. THOMAS (Ed.), *Proceedings of the 24th International Conference on Technology in Mathematics Education* (pp. 112–119). Auckland: Auckland Institute of Technology.

CAVANAGH, M., & MITCHELMORE, M. (2000b). Student misconceptions in interpreting basic graphic calculator displays. In T. NAKAHARA & M. KOYAMA (Eds.), *Proceedings of the Conference of the International Group for the Psychology of Mathematics Education* (Vol. 2, pp. 161–168). Hiroshima: PME.

CAVANAGH, M., & MITCHELMORE, M. (2000c). Students' technical difficulties in operating a graphics calculator. In J. BANA & A. CHAPMAN (Eds.), *Mathematics education beyond 2000: Proceedings of the Twenty-third Annual Conference of the Mathematics Education Research Group of Australasia (MERGA-23)* (pp. 142–148). Fremantle: MERGA.

CHAIKLIN, S. (2003). The Zone of Proximal Development in Vygotsky's Analysis of Learning and Instruction. In A. KOZULIN, B. GINDIS, V. S. AGEYEV, & S. M. MILLER (Eds.),*Vygotsky's Educational Theory in Cultural Context* (Chap. 2, pp. 39–64). Cambridge: Cambridge University Press.

CHEN, Z., & KLAHR, D. (1999). All Other Things Being Equal: Acquisition and Transfer of the Control of Variables Strategy. *Child Development*, *70*(5), 1098–1120

CHEUNG, A. C. K., & SLAVIN, R. E. (2013). The effectiveness of educational technology applications for enhancing mathematics achievement in K-12 classrooms: A meta-analysis. *Educational Research Review*, *9*, 88–113.

CLEMENT, J. (1985). Misconceptions in graphing. In L. STREEFLAND (Ed.), *Proceedings of the 9th Conference of the International Group for the Psychology of Mathematics Education* (Vol. 1, pp. 369–375). Noordwijkerhout: PME.

CONFREY, J., & SMITH, E. (1991). A Framework for Functions: Prototypes, Multiple Representations, and Transformations. In R. G. UNDERHILL (Ed.), *Proceedings of the 13th annual meeting of the North American Chapter of The International Group for the Psychology of Mathematics Education* (Vol. 1, pp. 57–63).

DE JONG, T. (2005). The Guided Discovery Principle in Multimedia Learning. In R. MAYER (Ed.), *The Cambridge Handbook of Multimedia Learning* (pp. 215–228). Cambridge: Cambridge University Press.

DE JONG, T., & LAZONDER, A. W. (2014). *The Guided Discovery Learning Principle in Multimedia Learning*. In R. MAYER (Ed.), (2nd ed., Chap. 15, pp. 371–390). Cambridge: Cambridge University Press.

DE JONG, T., & NJOO, M. (1992). Learning and Instruction with Computer Simulations: Learning Processes Involved. *Computer-Based Learning Environments and Problem Solving*, 411–427.

DE JONG, T., & VAN JOOLINGEN, W. R. (1998). Scientific Discovery Learning with Computer Simulations of conceptual domains. *Review of Educational Research*, 68(2), 179–201.

DOORMAN, M., DRIJVERS, P., GRAVEMEIJER, K., BOON, P., & REED, H. (2012). Tool use and the development of the function concept: from repeated calculations to functional thinking. *International Journal of Science and Mathematics Education*, 10(6), 1243–1267.

DÖRING, N., & BORTZ, J. (2016). *Forschungsmethoden und Evaluation in den Sozial- und Humanwissenschaften* (5th ed.). Berlin: Springer.

DREYFUS, T., & EISENBERG, T. (1987). On the Deep Structure of Functions. In J. C. BERGERON, N. HERSCOVICS, & C. KIERAN (Eds.), *Proceedings of the 11th International Conference of the International Group on Psychology of Mathematics Education* (Vol. 1, pp. 190–196). Montreal.

DRIJVERS, P. (2000). Students Encountering Obstacles Using a CAS. *International Journal of Computers for Mathematical Learning*, 5(3), 189–209.

DRIJVERS, P. (2001). The concept of parameter in a computer algebra environment. In M. VAN DEN HEUVEL-PANHUIZEN (Ed.), *Proceedings of the 25th Conference of the In ternational Group for the Psychology of Mathematics Education* (Vol. 2, pp. 385–392). Freudenthal Institute. Utrecht, Netherlands.

DRIJVERS, P. (2018). Empirical Evidence for Benefit? Reviewing Quantitative Research on the Use of Digital Tools in Mathematics Education. In L. BALL, P. DRIJVERS, S. LADEL, H.-S. SILLER, M. TABACH, & C. VALE (Eds.), *Uses of Technology in Primary and Secondary Mathematics Education* (pp. 161–175). ICME-13 Monographs. Cham: Springer International Publishing.

DRIJVERS, P. H. M. (2003). *Learning algebra in a computer algebra environment: Design research on the understanding of the concept of parameter*. Utrecht: Utrecht University.

DRIJVERS, P. H. M. (2004). Learning algebra in a computer algebra environment. *International Journal for Technology in Mathematics Education*, 11(3), 77–89.

DRIJVERS, P., BALL, L., BARZEL, B., HEID, M. K., CAO, Y., & MASCHIETTO, M. (2016). *Uses of technology in lower secondary mathematics education: A concise topical survey*. Springer open.

DRIJVERS, P., DOORMAN, M., BOON, P., REED, H., & GRAVEMEIJER, K. (2010). The teacher and the tool: instrumental orchestrations in the technology-rich mathematics classroom. *Educational Studies in Mathematics*, 75(2), 763–774.

DRIJVERS, P., & TROUCHE, L. (2008). From artifact to instrument: A theoretical Framework Behind the Orchestra Metaphor. In G. W. BLUME & M. K. HEID (Eds.), *Research on technology and the teaching and learning of mathematics: Vol. 2. Cases and Perspectives* (pp. 363–391). Information Age.

DUBINSKY, E., & HAREL, G. (1992). The nature of the process conception of function. In G. HAREL & E. DUBINSKY (Eds.), *The concept of function: Aspects of epistemology and pedagogy* (pp. 85–106). Washington: Mathematical Association of America.

DUNCAN, A. G. (2010). Teachers' views on dynamically linked multiple representations, pedagogical practices and students' understanding of mathematics using TI-Nspire in Scottish secondary schools. *ZDM: The International Journal on Mathematics Education, 42*(7), 763–774.

DUVAL, R. (1999). Representation, vision and visualization: Cognitive functions in mathematical thinking: Basic issues for learning. In F. HITT, & M. SANTOS (Eds.), *Proceedings of the 21st annual meeting of the North American Chapter of the International Group for the Psychology of Mathematics Education* (Vol. 1, pp. 3–36).

DUVAL, R. (2006). A cognitive analysis of problems of comprehension in a learning of mathematics. *Educational Studies in Mathematics, 61*(1-2), 103–131.

DUVAL, R. (2014). Commentary: Linking epistemology and semio-cognitive modeling in visualization. *ZDM, 46*(1), 159–170.

DUVAL, R. (2017). *Understanding the Mathematical Way of Thinking – The Registers of Semiotic Representations*. Springer International Publishing.

EBERS, P., PETERS- DASDEMIR, J., THURM, D., & WAGENER, O. (2019). Der Herausforderung der Digitalisierung im Mathematikunterricht in Fortbildungen begegnen. In A. BÜCHTER, M. GLADE, R. HEROLD-Btextsclasius, M. KLINGER, F. SCHACHT, & P. SCHERER (Eds.), *Vielfältige Zugänge zum Mathematikunterricht* (pp. 281–294). Wiesbaden: Springer Spektrum.

EISENBERG, T., & DREYFUS, T. (1994). On Understanding How Students Learn to Visualize Function Transformations. In E. DUBINSKY, A. H. SCHOENFELD, & J. KAPUT, (Eds.), *Research in collegiate mathematics education* (Vol. 1, pp. 45–68). Providence, RI: American Mathematical Society.

ELLIS, A. B., & GRINSTEAD, P. (2008). Hidden lessons: How a focus on slope-like properties of quadratic functions encouraged unexpected generalizations. *Journal of Mathematical Behavior, 27*(4), 277–296.

FEIERABEND, S., PLANKENHORN, T., & RATHGEB, T. (2016). *JIM 2016 Jugend, Information, (Multi-) Media*. Medienpädagogischer Forschungsverbund Südwest (mpfs).

FEIERABEND, S., RATHGEB, T., & REUTTER, T. (2018). *JIM 2018 Jugend, Information, Medien – Basisuntersuchung zum Medienumgang 12- bis 19-Jähriger*. Stuttgart: Medienpädagogischer Forschungsverbund Südwest (mpfs).

FERRARA, F., PRATT, D., & ROBUTTI, O. (2006). The role and uses of technologies for the teaching of algebra and calculus. In A. GUTIÉRREZ & P. BOERO (Eds.), *Handbook of Research on the Psychology of Mathematics Education: Past, Present, Future* (pp. 237–273). Rotterdam: Sense.

FOY, P., ARORA, A., & STANCO, G. M. (Eds.). (2013). *TIMSS 2011 User Guide for the International Database Released Items Mathematics – Eighth Grade*. TIMSS & PIRLS International Study Center.

FREUDENTHAL, H. (1962). Logical analysis and critical survey. In H. FREUDENTHAL (Ed.), *Report on the relations between arithmetic and algebra* (pp. 20–41). Groningen: J.B. Wolters.

FREUDENTHAL, H. (1973). *Mathematik als Pädagogische Aufgabe.* Stuttgart: Ernst Klett Verlag.

FREUDENTHAL, H. (1983). *Didactical phenomenology of mathematical structures.* New York: Kluwer.

FUJII, T. (2003). Probing students' understanding of variables through cognitive conflict problems: is the concept of a variable so difficult for students to understand? In N. A. PATEMAN, B. J. DOUGHERTY, & J. T. ZILLIOX (Eds.), *Proceedings of the 27th International Conference for the Psychology of Mathematics Education* (Vol. 1, pp. 49–65). Hawai'i: PME27 & PME-NA25.

Fünfundvierzig. (2011). Retrieved from https://www.iqb.hu-berlin.de/vera/aufgaben/ma1/system/taskpool/getTaskFile?fileid=5651

FURINGHETTI, F., & PAOLA, D. (1994). Parameters, unknowns and variables: A little difference? In J. P. DA PONTE, & J. F. MATOS (Eds.), *Proceedings of the 18th Conference of the International Group for the Psychology of Mathematics Education* (Vol. 2, pp. 368–375). Lissabon: PME.

GADOWSKY, K. D. M. (2001). *A window on learning by inquiry with technology: What meanings do grade 11 students construct when exploring transformations on functions using graphing technology?* (Master's thesis, Simon Fraser University).

GERSEMEHL, I., JÖRGENS, T., JÜRGENSEN- ENGL, T., RIEMER, W., SONNTAG, R., & SPIELMANS, H. (2013). *Lambacher Schweizer 9: Mathematik für Gymnasien / Nordrhein-Westfalen* (2nd ed.). Stuttgart: Klett.

GERVER, R. K., & SGROI, R. J. (2003). Creating and Using Guided-Discovey Lessons. *The Mathematics Teacher, 96*(1), 6–13.

GLENN, J. A., & LITTLER, G. H. (Eds.). (1984). *Dictionary of Mathematics.* Totowa, N.J.: Barnes & Noble.

GÖBEL, L. (2017). Vergleich verschiedener Visualisierungen anhand von Schülerdokumenten bei der Konzeptualisierung von Parametern bei quadratischen Funktionen. In U. KORTENKAMP & A. KUZLE (Eds.), *Beiträge zum Mathematikunterricht 2017* (Vol. 1, pp. 322–329). Münster: WTM.

GÖBEL, L. (2018). "Power of Speed" oder "Discovery by Slowness": Technologiegestütztes Guided Discovery bei der Konzeptualisierung von Parametern bei quadratischen Funktionen. In FACHGRUPPE DIDAKTIK DER MATHEMATIK DER UNIVERSITÄT PADERBORN (Ed.), *Beiträge zum Mathematikunterricht 2018* (pp. 623–626). Münster: WTM-Verlag.

GÖBEL, L. (2019). Dynamisch vs. statisch! Verschiedene Visualisierungen bei der Konzeptualisierung von Parametern quadratischer Funktionen. In A. FRANK, S. KRAUSS, & K. BINDER (Eds.), *Beiträge zum Mathematikunterricht 2019* (p. 1395). Münster: WTM-Verlag.

GÖBEL, L., & BARZEL, B. (2016). Vergleich verschiedener dynamischer Visualisierungen zur Konzeptualisierung von Parametern bei quadratischen Funktionen. In INSTITUT FÜR MATHEMATIK UND INFORMATIK HEIDELBERG (Ed.), *Beiträge zum Mathematikunterricht 2016: Vorträge auf der 50. Tagung für Didaktik der Mathematik vom 07.03.2016 bis 11.03.2016 in Heidelberg* (pp. 313–316). Münster: WTM.

GÖBEL, L., & BARZEL, B. (2019). Dynamic vs. static! Different visualisations to conceptualize parameters of quadratic functions. In U. T. JANKVIST, M. VAN DEN HEUVEL-PANHUIZEN, & M. VELDHUIS (Eds.), *Proceedings of the Eleventh Congress of the European Society for Research in Mathematics Education (CERME11)* (pp. 2836–2837). Proceedings of the Eleventh Congress of the European Society for Research in Mathematics Education

(CERME11). Utrecht University. Utrecht, Netherlands: Freudenthal Group &, Freudenthal Institute, Utrecht University, & ERME. Retrieved from https://hal.archives-ouvertes.fr/hal-02428234254

GÖBEL, L., BARZEL, B., & BALL, L. (2017). "Power of Speed" or "Discovery by Slowness": Technology-assisted Guided Discovery to Investigate the Role of Parameters in Quadratic Functions. In G. ALDON & J. TRGALOVA (Eds.), *Proceedings of the 13th International Conference on Technology in Mathematics Teaching* (pp. 113–123). France.

GOLDENBERG, E. P. (1988). Mathematics, metaphors, and human factors: Mathematical, technical, and pedagogical challenges in the educational use of graphical representation of functions. *Journal of Mathematical Behavior, 7*(2), 135–173.

GOLDENBERG, E. P., & KLIMAN, M. (1988). *Metaphors for understanding graphs: What you see is what you see*. Retrieved from http://eric.ed.gov/?id=ED303369

GÖSSLING, J. M. (2010). *Selbständig entdeckendes Experimentieren Lernwirksamkeit der Strategieanwendung* (Doctoral dissertation, Universität Duisburg-Essen).

GREEFRATH, G., OLDENBURG, R., SILLER, H.-S., ULM, V., & WEIGAND, H.-G. (2016). *Didaktik der Analysis: Aspekte und Grundvorstellungen zentraler Begriffe*. Berlin: Springer Spektrum.

GUIN, D., & TROUCHE, L. (1999). The Complex Process of Converting Tools into Mathematical Instruments: The Case of Calculators. *International Journal of Computers for Mathematical Learning, 3*, 195–227.

HARRIS, J. (1710). *Lexicon Technicum, or, an Universal English Dictionary of Arts and Sciences: Explaining Not Only the Terms of Art, but the Arts Themselves*. London: D. Brown.

HART, W. W. (1951). *A Second Course in Algebra* (2nd ed., enlarged). Boston: D. C. Heath & CO.

HECK, A. (2001). Variables in Computer Algebra, Mathematics and Science. *International Journal of Computer Algebra in Mathematics Education, 8*(3), 195–221.

HEID, M. K. (2018). Digital Tools in Lower Secondary School Mathematics Education: A Review of Qualitative Research on Mathematics Learning of Lower Secondary School Students. In L. BALL, P. DRIJVERS, S. LADEL, H.-S. SILLER, M. TABACH, & C. VALE (Eds.), *Uses of Technology in Primary and Secondary Mathematics Education* (pp. 177–201). ICME-13 Monographs. Cham: Springer International Publishing.

HERMANN, G. (1969). Learning by Discovery. *The Journal of Experimental Education, 38*(1), 58–72.

HEROLD-BLASIUS, R. (2019). *Problemlösen mit Strategieschlüsseln. Eine explorative Studie zur Unterstützung von Problembearbeitungsprozessen bei Dritt- und Viertklässlern*. (Doctoral dissertation, Universität Duisburg-Essen).

HEUGL, H., KINGER, W., & LECHNER, J. (1996). *Mathematikunterricht mit Computeralgebra-Systemen: Ein didaktisches Lehrbuch mit Erfahrungen aus dem österreichischen DERIVE-Projekt*. Bonn: Addison-Wesley.

HIEBERT, J., & CARPENTER, T. P. (1992). Learning and teaching with understanding. In D. A. GROUWS (Ed.), *Handbook of research on mathematics teaching and learning: A project of the National Council of Teachers of Mathematics* (Chap. 4, pp. 65–97). New York: Macmillan.

HIELE, P. M. V. (1986). *Structure and Insight: A Theory of Mathematics Education*. Developmental Psychology Series. Orlando: Academic Press.

HIRSCH, C. R. (1977). The Effects of Guided Discovery and Individualized Instructional Packages on Initial Learning, Transfer, and Retention in Second-Year Algebra. *Journal for Research in Mathematics Education*, 8(5), 359.

HÖNEKE, S. (2019). *Analyse des Technologieeinsatzes im Rahmen einer guided discovery zur Konzeptualisierung von Parametern bei quadratischen Funktionen* (unpublished master's thesis, Universität Duisburg-Essen).

JANVIER, C. (1978). *The interpretation of complex cartesian graphs representing situations: Studies and teaching experiments* (Dissertation, Nottingham University, Nottingham, Vereinigtes Königreich).

KAPUT, J. J. (1985). Representation and problem solving: Methodological issues related to modeling. In E. A. SILVER (Ed.), *Teaching and learning mathematical problem solving: Multiple research perspectives* (pp. 298–381). Hillsdale: Erlbaum.

KAPUT, J. J. (1989). Linking representations in the symbol systems of algebra. In S. WAGNER & C. KIERAN (Eds.), *Research issues in the learning and teaching of algebra* (pp. 167–194). Hillsdale: Erlbaum.

KAPUT, J. J. (1992). Technology and Mathematics Education. In D. GROUWS (Ed.), *Handbook of Research on Mathematics Teaching and Learning* (pp. 515–556). NCTM.

KERSH, B. Y. (1962). The motivating effect of learning by directed discovery. *Journal of Educational Psychology*, 53(2), 65–71.

KIERAN, C. (1992). The Learning and Teaching of School Algebra. In D. A. GROUWS (Ed.), *Handbook of Research on Mathematics Teaching and Learning* (pp. 390–419). New York: NCTM.

KIERAN, C., & DRIJVERS, P. (2006). The Co-Emergence of Machine Techniques, Paper-and-Pencil Techniques, and Theoretical Reflection: A Study of CAS Use in Secondary School Algebra. *International Journal of Computers for Mathematical Learning*, 11(2), 205–263.

KIMANI, P. M. (2008). *Calculus students' understandings of the concepts of function transformation, function composition, function inverse, and the relationships among the three concepts* (Doctoral dissertation, Syracuse University).

KIRSCHNER, P. A., SWELLER, J., & CLARK, R. E. (2006). Why Minimal Guidance During Instruction Does Not Work: An Analysis of the Failure of Constructivist, Discovery, Problem-Based, Experiential, and Inquiry-Based Teaching. *Educational Psychologist*, 41(2), 75–86.

KLEIN ALTSTEDDE, J. (2016). *Probleme und Potential des Zugmodus im Rahmen einer explorativen Lernumgebung zur Bedeutung von Parametern bei quadratischen Funktionen* (unpublished state exam thesis, Universität Duisburg-Essen).

KLINGER, M. (2018). *Funktionales Denken beim übergang von der Funktionenlehre zur Analysis*. Springer Spektrum.

KLINGER, M. (2019). Grundvorstellungen versus Concept Image? Gemeinsamkeiten und Unterschiede beider Theorien am Beispiel des Funktionsbegriffs. In A. BÜCHTER, M. GLADE, R. HEROLD-BLASIUS, M. KLINGER, F. SCHACHT, & P. SCHERER (Eds.), *Vielfältige Zugänge zum Mathematikunterricht* (Chap. 5, pp. 61–75). Wiesbaden: Springer Spektrum.

KMK (SEKRETARIAT DER STÄNDIGEN KONFERENZ DER KULTUSMINISTER DER LÄNDER DER BUNDESREPUBLIK DEUTSCHLAND) (Ed.). (2004). *Bildungsstandards im Fach Mathematik für den Mittleren Schulabschluss: Beschluss vom 4.12.2003*. München: Kluwer.

KMK (SEKRETARIAT DER STÄNDIGEN KONFERENZ DER KULTUSMINISTER DER LÄNDER DER BUNDESREPUBLIK DEUTSCHLAND) (Ed.). (2015). *Bildungsstandards im Fach Math-*

ematik für die Allgemeine Hochschulreife (Beschluss der Kultusministerkonferenz vom 18.10.2012). Köln: Kluwer.

KMK (SEKRETARIAT DER STÄNDIGEN KONFERENZ DER KULTUSMINISTER DER LÄNDER DER BUNDESREPUBLIK DEUTSCHLAND) (Ed.). (2016). *Bildung in der digitalen Welt: Strategie der Kultusministerkonferenz*. Berlin.

KRÜGER, K. (2002). Funktionales Denken – „alte" Ideen & „neue" Medien. In W. HERGET, R. SOMMER, H.-G. WEIGAND, & T. WETH (Eds.), *Medien verbreiten Mathematik: Bericht über die 19. Arbeitstagung des Arbeitskreises „Mathematikunterricht und Informatik" in der Gesellschaft für Didaktik der Mathematik e. V.* (pp. 120–127). Hildesheim: Franzbecker.

KRÜGER, K. (2019). Functional Thinking: The History of a Didactical Principle. In H.-G. WEIGAND, W. MCCALLUM, M. MENGHINI, M. NEUBRAND, & G. SCHUBRING (Eds.), *The Legacy of Felix Klein* (pp. 35–53). Cham: Springer International Publishing.

KÜCHEMANN, D. E. (1981). Algebra. In K. M. HART (Ed.), *Children's understanding of mathematics: 11–16* (pp. 102–119). London: Murray.

KURTUL, S. (2018). *Kategorisierung von Schülerantworten zu der Bedeutung des Parameters c bei quadratischen Funktionen* (unpublished master's thesis, Universität Duisburg-Essen).

LAZONDER, A. W., & HARMSEN, R. (2016). Meta-Analysis of Inquiry-Based Learning. *Review of Educational Research, 86*(3), 681–718.

LEINHARDT, G., ZASLAVSKY, O., & STEIN, M. K. (1990). Functions, graphs, and graphing: Tasks, learning, and teaching. *Review of Educational Research, 60*(1), 1–64.

LI, Q., & MA, X. (2010). A Meta-analysis of the Effects of Computer technology on School Students' Mathematics Learning. *Educational Psychology Review, 22*(3), 215–243.

LINDMEIER, A. (2018). Innovation durch digitale Medien im Fachunterricht? – Ein Forschungsüberblick aus fachdidaktischer Perspektive. In M. ROPOHL, A. LINDMEIER, H. HÄRTIG, L. KAMPSCHULTE, A. MÜHLING, & J. SCHWANEWEDEL (Eds.), *Medieneinsatz im mathematisch-naturwissenschaftlichen Unterricht. Fachübergreifende Perspetiven auf zentrale Fragestellungen* (Chap. 3, pp. 55–97). Hamburg: Joachim Herz Verlag.

LORENZ, R., BOS, W., ENDBERG, M., EICKELMANN, B., GRAFE, S., & VAHRENHOLD, J. (Eds.). (2017, December 8). *Schule digital – der Länderindikator 2017*. Münster: Waxmann Verlag GmbH. Retrieved from https://www.ebook.de/de/product/30819706/schule_digital_der_laenderindikator_2017.html

MACGREGOR, M., & STACEY, K. (1997). Students' understanding of algebraic notation: 11–15. *Educational Studies in Mathematics, 33*, 1–19.

MALLE, G. (1993). *Didaktische Probleme der elementaren Algebra*.

MALLE, G. (2000). Zwei Aspekte von Funktionen: Zuordnung und Kovariation. *Mathematik Lehren, Heft 103*, 8–11.

MAYER, R. E. (2004). Should There Be a Three-Strikes Rule Against Pure Discovery Learning? *American Psychologist, 59*(1), 14–19.

MAYER, R. E. (2009). Constructivism as a Theory of Learning Versus Constructivism as a Prescription for Instruction. In S. TOBIAS & T. M. DUFFY (Eds.), *Constructivist Instruction: Success or failure?* (Chap. 10, pp. 184–200). Routledge Taylor & Francis Group.

MAYRING, P. (2004). Qualitative Content Analysis. In U. FLICK, E. VON KARDORFF, & I. STEINKE (Eds.), (Chap. 5.12, pp. 266–269). London: SAGE.

MAYRING, P. (2015). *Qualitative Inhaltsanalyse: Grundlagen und Techniken* (12th ed.). Weinheim: Beltz.

MAYRING, P. (2016). *Einführung in die qualitative Sozialforschung* (6th ed.). Weinheim: Beltz.

MCCLARAN, R. R. (2013). *Investigating the impact of interactive applets on students' understanding of parameter changes to parent functions: an explanatory mixed methods study* (Doctoral dissertation, University of Kentucky).

METCALF, R. C. (2007). *The nature of students' understanding of quadratic functions* (Doctoral dissertation, State University of New York, Buffalo).

MITCHELMORE, M., & CAVANAGH, M. (2000). Students' difficulties in operating a graphics calculator. *Mathematics Education Research Journal, 12*(3), 254–268.

MONAGHAN, J., TROUCHE, L., & BORWEIN, J. M. (Eds.). (2016). *Tools and mathematics: Instruments for learning.* Cham: Springer.

MOSKALENKO, S. (2019). *Analyse der Parameterrollen im Rahmen einer Guided Discovery im Bereich quadratischer Funktionen* (unpublished master's thesis, Universität Duisburg-Essen).

MOSSTON, M. (1972). *Teaching: From Command to Discovery.* Belmont, California: Wadsworth.

MOSSTON, M., & ASHWORTH, S. (2002). *Teaching Physical Education* (5th ed.). San Francisco: Cummings.

MSB NRW (MINISTERIUM FÜR SCHULE UND BILDUNG DES LANDES NORDRHEIN-WESTFALEN) (Ed.). (2019). *Kernlehrplan für die Sekundarstufe I Gymnasium in Nordrhein-Westfalen: Mathematik.* Düsseldorf.

MSW NRW (MINISTERIUM FÜR SCHULE UND WEITERBILDUNG DES LANDES NORDRHEIN-WESTFALEN) (Ed.). (2007). *Kernlehrplan für das Gymnasium – Sekundarstufe I (G8) in Nordrhein-Westfalen: Mathematik.* Frechen: Ritterbach.

MSW NRW (MINISTERIUM FÜR SCHULE UND WEITERBILDUNG DES LANDES NORDRHEIN-WESTFALEN). (2012).*Gebrauch von graphikfähigen Taschenrechnern im Mathematikunterricht der gymnasialen Oberstufe und des Beruflichen Gymnasiums: RdErl. d. Ministeriums für Schule und Weiterbildung v. 27.06.2012 (523-6.08.01-105571).* Retrieved from http://www.standardsicherung.schulministerium.nrw.de/cms/zentralabitur-gost/faecher/getfile.php?file=3347

NABERS, L. (2016). *Vergleich unterschiedlicher dynamischer Visualisierungen anhand von Schülerdokumenten nach einer Erkundung zur Parameterbedeutung von quadratischen Funktionen* (unpublished master's thesis, Universität Duisburg-Essen).

OECD (ORGANISATION FOR ECONOMIC CO- OPERATION AND DEVELOPMENT) (Ed.). (2015). *Students, computers and learning: Making the connection.* Paris: OECD.

PALMER, S. E. (1978). Fundamental aspects of cognitive representation. In E. ROSCH & B. LLOYD (Eds.), *Cognition and categorization* (pp. 259–303). Hillsdale: Lawrence Erlbaum.

PENGLASE, M., & ARNOLD, S. (1996). The Graphics Calculator in Mathematics Education: A Critical Review of Recent Research. *Mathematics Education Research Journal, 8*(1), 58–90.

PHILIPP, K. (2012). *Experimentelles Denken: Theoretische und empirische Konkretisierung einer mathematischen Kompetenz.* Wiesbaden: Springer Spektrum.

PIHLAP, S. (2017). The Impact of Computer use on Learning of Quadratic Functions. *International Journal for Technology in Mathematics Education, 24*(2), 59–66.

PINKERNELL, G. (2015). Reasoning with dynamically linked multiple representations of functions. In K. KRAINER & N. VONDROVÁ (Eds.), *CERME 9 – Ninth Congress of the European Society for Research in Mathematics Education* (pp. 2531–2537).

PINKERNELL, G., & VOGEL, M. (2016). Zum Einsatz softwarebasierter multipler Repräsentationen von Funktionen im Mathematikunterricht. In G. HEINTZ, G. PINKERNELL, & F. SCHACHT (Eds.), *Digitale Werkzeuge für den Mathematikunterricht* (pp. 231–242). MNU – Verband zur Förderung des MINT-Unterrichts.

PINKERNELL, G., & VOGEL, M. (2017). "Das sieht aber anders aus" – zu Wahrnehmungsfallen beim Unterricht mit computergestützten Funktionsdarstellungen. *Der Mathematikunterricht, 6*, 38–46.

RADFORD, L. (2014). Towards an embodied, cultural, and material conception of mathematics cognition. *ZDM, 46*(3), 349–361.

RAKES, C. R., VLENTINE, J. C., MCGATHA, M. B., & RONAU, R. N. (2010). Methods of Instructional Improvement in Algebra. *Review of Educational Research, 80*(3), 372–400.

RASCH, B., FRIESE, M., HOFMANN, W., & NAUMANN, E. (2014). *Quantitative Methoden 2: Einführung in die Statistik für Psychologen und Sozialwissenschaftler*. Berlin: Springer.

REID, D., ZHANG, J., & CHEN, Q. (2003). Supporting scientific discovery learning in a simulation environment. *Journal of Computer Assisted Learning, 19*(1), 9–20.

RONAU, R. N., RAKES, C. R., BUSH, S. B., DRISKELL, S. O., NIESS, M. L., & PUGALEE, D. K. (2014). A Survey of Mathematics Education Technology Dissertation Scope and Quality: 1968–2009. *American Educational Research Journal, 51*(5), 974–1006.

ROSNICK, P. (1981). Some Misconceptions concerning the concept of variable. *The Mathematics Teacher, 74*(6), 418–420.

RUCHNIEWICZ, H., & BARZEL, B. (2019). Technology Supporting Student Self-Assessment in the Field of Functions-A Design-Based Research Study. In G. ALDON & J. TRGALOVA (Eds.), *Technology in Mathematics Teaching – Selected Papers of the 13th ICTMT Conference* (Vol. 13, pp. 49–74). Mathematics Education in the Digital Era. Springer International Publishing.

RUCHNIEWICZ, H., & GÖBEL, L. (2019). Wie digitale Medien funktionales Denken unterstützen können – Zwei Beispiele. In A. BÜCHTER, M. GLADE, R. HEROLD-BLASIUS, M. KLINGER, F. SCHACHT, & P. SCHERER (Eds.), *Vielfältige Zugänge zum Mathematikunterricht* (pp. 249–262). Wiesbaden: Springer Spektrum.

SCHMIDT-THIEME, B., & WEIGAND, H.-G. (2015). Medien. In R. BRUDER, L. HEFENDEHL-HEBEKER, B. SCHMIDT-THIEME, & H.-G. WEIGAND (Eds.), *Handbuch der Mathematikdidaktik* (pp. 461–490). Berlin: Springer Spektrum.

SCHOENFELD, A. H., & ARCAVI, A. (1988). On the Meaning of Variable. *The Mathematics Teacher, 81*(6), 420–427.

SCHUBRING, G. (2007). Der Aufbruch zum „funktionalen Denken": Geschichte des Mathematikunterrichts im Kaiserreich: 100 Jahre Meraner Reform. *NTM International Journal of History and Ethics of Natural Sciences, Technology and Medicine, 15*(1), 1–17.

SEVER, G., & YERUSHALMY, M. (2007). To Sense and to Visualize Functions: The Case of Graphs' Stretching. In D. PITTA-PANTAZI & G. PHILIPPOU (Eds.), *European Research in Mathematics Education V: Proceeedings of the Fifth Congress of the European Society for Research in Mathematics Education (CERME 5, February 22-26,2007)* (pp. 1509–1518). University of Cyprus & ERME. Larnaca, Cyprus.

SFARD, A. (1991). On the dual nature of mathematical conceptions: Reflections on processes and objects as different sides of the same coin. *Educational Studies in Mathematics*, 22(1), 1–36.

SIEBEL, F. (2005). *Elementare Algebra und ihre Fachsprache – Eine allgemein-mathematische Untersuchung* (Doctoral dissertation, Technische Universität Darmstadt).

SINCLAIR, N. (2004). Computer-Based Technologies and Plausible Reasoning. In M. P. CARLSON & C. RASMUSSEN (Eds.), *Making the Connection* (pp. 233–244). The Mathematical Association of America.

SINCLAIR, N., & YERUSHALMY, M. (2016). Digital Technology in Mathematics Teaching and Learning. In Á. GUTIÉRREZ, G. C. LEDER, & P. BOERO (Eds.), *The Second Handbook of Research on the Psychology of Mathematics Education* (pp. 235–274). Rotterdam: Sense Publishers.

SOKOLOWSKI, A., LI, Y., & WILLSON, V. (2015). The effects of using exploratory computerized environments in grades 1 to 8 mathematics: a meta-analysis of research. *International Journal of STEM Education*, 2(1).

SPECHT, B. J. (2009). *Variablenverständnis und Variablen verstehen – Empirische Untersuchungen zum Eifnluss sprachlicher Formulierungen in der Primar- und Sekundarstufe.* Hildesheim, Berlin: Franzbecker.

STEENBERGEN-HU, S., & COOPER, H. (2013). A meta-analysis of the effectiveness of intelligent tutoring systems on K-12 students' mathematical learning. *Journal of Educational Psychology*, 105(4), 970–987.

TALL, D., SMITH, D., & PIEZ, C. (2008). Technology and calculus. In M. K. HEID & G. W. BLUM (Eds.), *Research on technology and the teaching and learning of mathematics: Research syntheses* (Chap. 5, Vol. 1, pp. 207–258). Charlotte: Information Age.

TALL, D., & VINNER, S. (1981). Concept image and concept definition in mathematics with particular reference to limits and continuity. *Educational Studies in Mathematics*, 12(2), 151–169.

TERFURTH, S. (2016). *Analyse von Merkblättern von Lernenden hinsichtlich Sprache, Struktur und Inhalt im Bereich der Parameterdeutung bei quadratischen Funktionen* (unpublished master's thesis, Universität Duisburg-Essen).

THOMAS, M. O. J. (2008). Conceptual representations and versatile mathematical thinking. In M. NISS (Ed.), *ICMI-10 proceedings and regular lectures*, Copenhagen, Denmark.

THURM, D. (2020). *Digitale Werkzeuge im Mathematikunterricht integrieren.* Wiesbaden: Springer Spektrum.

THURM, D., & BARZEL, B. (2020). Effects of a professional development program for teaching mathematics with technology on teachers' beliefs, self-efficacy and practices. *ZDM*

TROPFKE, J. (1902). *Geschichte der Elementar-Mathematik in systematischer Darstellung.* Leipzig: von Veit.

TROUCHE, L. (2004). Managing the Complexity of Human/Machine Interactions in Computerized Learning Environments: Guiding Students' Command Process through Instrumental Orchestrations. *International Journal of Computers for Mathematical Learning*, 9(3), 281–307.

TROUCHE, L. (2005a). An instrumental approach to mathematics learning in symbolic calculator environments. In D. GUIN, K. RUTHVEN, & L. TROUCHE (Eds.), *The Didactical Challenge of Symbolic Calculators: Turning a Computer into a Mathematical Instrument* (Chap. 6, pp. 137–162). New York, USA: Springer.

TROUCHE, L. (2005b). Instrumental genesis, individual and social aspects. In D. GUIN, K. RUTHVEN, & L. TROUCHE (Eds.), *The Didactical Challenge of Symbolic Calculators: Turning a Computer into a Mathematical Instrument* (Chap. 8, pp. 197–230). New York, USA: Springer.

TROUCHE, L. (2014). Instrumentation in Mathematics education. In S. LERMAN (Ed.), *Encyclopedia of Mathematics Education* (pp. 307–313). Springer Netherlands.

TROUCHE, L., & DRIJVERS, P. (2010). Handheld technology for mathematics education: flashback into the future. *ZDM: The International Journal on Mathematics Education, 42*(7), 667–681.

URSINI, S., & TRIGUEROS, M. (2001). A model for the uses of variable in elementary algebra. In M. v. d. HEUVEL-PANHUIZEN (Ed.), *Proceedings of the 25th Conference of the International Group for the Psychology of Mathematics Education* (Vol. 4, pp. 327–334). Freudenthal Institute. Utrecht, Netherlands: PME.

URSINI, S., & TRIGUEROS, M. (2004). How do high school students interpret parameters in algebra? In M. J. HOINES & A. B. FUGLESTAD (Eds.), *Proceedings of the 28th Conference of the International Group for the Psychology of Mathematics Education* (Vol. 4, pp. 361–368). Bergen: PME.

USISKIN, Z. (1988). Conceptions of school algebra and uses of variable. In A. F. COXFORD (Ed.), *The ideas of Algebra, K-12 (1988 Yearbook of the NCTM)* (pp. 8–19). Reston: NCTM.

VAN DE GIESSEN, C. (2002). The visualisation of paramters. In M. BOROVCNIK & H. KAUTSCHITSCH (Eds.), *Technology in mathematics teaching. Proceedings of ICTMT5* (pp. 97–100). Oebv&hpt Verlagsgesellschaft. Vienna.

VERILLON, P., & RABARDEL, P. (1995). Cognition and Artifacts: A Contribution to the Study of Though in Relation to Instrumented Activity. *European Journal of Psychology of Educatio, 10*(1), 77–101.

VINNER, S., & DREYFUS, T. (1989). Images and definitions for the concept of function. *Journal of Research in Mathematics Education, 20*(4), 356–366.

VINNER, S., & HERSHKOWITZ, R. (1980). Concept images and common cognitive paths in the development of some simple geometrical concepts. In *Proceedings of the fourth international conference for the psychology of mathematics education* (pp. 177–184). Berkeley: University of California.

VOLLRATH, H.-J. (1989). Funktionales Denken. *Journal für Mathematik-Didaktik, 10*(1), 3–37.

VOLLRATH, H.-J., & ROTH, J. (2012). *Grundlagen des Mathematikunterrichts in der Sekundarstufe* (2. Auflage). Heidelberg: Spektrum.

VOM HOFE, R. (1992). Grundvorstellungen mathematischer Inhalte als didaktisches Modell. *Journal für Mathematik-Didaktik, 13*(4), 345–364.

VOM HOFE, R. (1995). *Grundvorstellungen mathematischer Inhalte*. Heidelberg: Spektrum.

VOM HOFE, R. (2003). Grundbildung durch Grundvorstellungen. *Mathematik Lehren, 118*, 4–8.

VOM HOFE, R., & BLUM, W. (2016). „Grundvorstellungen" as a category of subject-matter didactics. *Journal für Mathematik-Didaktik, 37*(1), 225–254.

VOM HOFE, R., KLEINE, M., WARTHA, S., BLUM, W., & PEKRUN, R. (2005). On the Role of „Grundvorstellungen" for the Development of Mathematical Literacy – First Results of the Longitudinal Study PALMA. *Mediterranean Journal for Research in Mathematics Education, 4*(2), 67–84.

VYGOTSKY, L. S. (1978). *Mind in Society: The Development of Higher Psychological Processes.* Harvard University Press.

WAGNER, S. (1981). Conservation of equation and function under transformations of variable. *Journal for Research in Mathematics Education, 12*(2), 107–118.

WAGNER, S. (1983). What are These Things Called Variables. *The Mathematics Teacher, 76*(7), 474–479.

WARREN, E. (1999). The concept of a variable; gauging students understanding. In O. ZASLAVSKY (Ed.), *Proceedings of the 23rd International Conference for the Psychology of Mathematics Education* (Vol. 4, pp. 313–320).

WESSELER, J. (2016). *Fallstudie zum Vorstellungsaufbau von Parametern bei quadratischen Funktionen in Abhängigkeit verschiedener dynamischer Visualisierungen* (unpublished master's thesis, Universität Duisburg-Essen).

WINTER, H. (1988). Lernen durch Entdecken? *Mathematik Lehren, 28*, 6–13.

WINTER, H. (1989). *Entdeckendes Lernen im Mathematikunterricht.* Braunschweig: Vieweg.

WINTER, H. W. (2016). *Entdeckendes Lernen im Mathematikunterricht: Einblicke in die Ideengeschichte und ihre Bedeutung für die Pädagogik* (3rd ed.). Wiesbaden: Springer Spektrum.

WITTMANN, G. (2008). *Elementare Funktionen und ihre Anwendungen.* Berlin: Spektrum.

ZASLAVSKY, O. (1997). Conceptual Obstacles in the Learning of Quadratic Functions. *Focus on Learning Problems in Mathematics, 19*(1), 20–44.

ZAZKIS, R., LILJEDAHL, P., & GADOWSKY, K. (2003). Conceptions of function translation: obstacles, intuitions, and rerouting. *The Journal of Mathematical Behavior, 22*(4), 435–448.

ZBIEK, R. M., HEID, M. K., BLUME, G. W., & DICK, T. P. (2007). Research on technology in mathematics education: A perspective of constructs. In F. K. LESTER (Ed.), *Second handbook of research on mathematics teaching and learning* (pp. 1169–1207). Charlotte: Information Age.

Printed in the United States
By Bookmasters